VARIATION IN ANIMALS AND PLANTS

VARIATION IN ANIMALS AND PLANTS

By

H. M. VERNON, M.A., M.D.

GYAN

Published by

GYAN

Gyan Publishing House
5, Ansari Road
Daryaganj, New Delhi-110002
Phone: 011-47034999, 9811692060
E-mail: books@gyanbooks.com

Distribution Network
gyanbooks.com
India, USA, Canada, UK, Australia, France

ISBN : 978-81-212-6476-1 (PB)
First Published, 1902

2nd Impression 2022

Printed at: Gyan Press, Delhi.

VARIATION IN ANIMALS AND PLANTS
Author: H. M. VERNON

VARIATION

IN ANIMALS AND PLANTS

BY

H. M. VERNON, M. A., M. D.

1902

PREFACE.

In this little book I have endeavoured to give a brief account of the subject of "Variation," so far as the present state of our knowledge admits. Though I did not in any degree aim at giving a complete representation of the subject, yet I hope that most of the more important and more recent work has been included. I have not treated Variation of Plants so fully as that of Animals, and from lack of thorough acquaintance with the literature, have probably made some omissions of real importance to the adequate comprehension of the subject in its bearing on living organisms taken as a whole. For such I offer my apologies. I have purposely avoided any higher mathematics in discussing the facts of variation, as it seemed out of place in a book of this character.

An obvious criticism upon the contents of the book will be that I have given greater prominence to my own researches than their intrinsic importance warrants. To this I frankly plead guilty, urging in extenuation that I did not intend to write a text-book in the ordinary sense of the term. Thus some of the hypotheses and interpretations of facts which I have given are my own personal opinions, and by no means current views held in general acceptation. Also a few of the data published in the latter part of Chapter VI., and those on "identical twins" in Chapter IV., are here published for the first time, and are for this reason given rather *in extenso.*

As regards the illustrations, I am indebted to the Royal Society for the blocks of Figs. 17, 18, 20, and 21, whilst Figs. 3, 6, 9, 16, 19, and 25 have been copied from illustrations in their publications. Figs. 22, 23, and 24 are copied from Davenport's "Experimental Morphology," and Figs. 7 and 8 from Bateson's "Materials for the Study of Variation." The rest are either original, or from sources indicated in the text.

I desire to take this opportunity of expressing my obligations to Mr. E. S. Goodrich for his kindness in reading over the manuscript, and to Professors W. F. R. Weldon and S. H. Vines, for their useful advice and suggestions.

CONTENTS.

PART I.

THE FACTS OF VARIATION.

CHAPTER I.

THE MEASUREMENT OF VARIATION.

CHAPTER II.

DIMORPHISM AND DISCONTINUOUS VARIATION.

CHAPTER III.

CORRELATED VARIATIONS.

PART II.

THE CAUSES OF VARIATION.

CHAPTER IV.

BLASTOGENIC VARIATIONS.

CHAPTER V.

BLASTOGENIC VARIATIONS (*Continued*).

CHAPTER VI.

CERTAIN LAWS OF VARIATION.

CHAPTER VII.

THE EFFECT OF TEMPERATURE AND OF LIGHT.

CHAPTER VIII.

THE EFFECT OF MOISTURE AND OF SALINITY.

CHAPTER IX.

THE EFFECT OF FOOD AND OF PRODUCTS OF METABOLISM.

CHAPTER X.

THE EFFECTS OF CONDITIONS OF LIFE IN GENERAL.

PART III.

VARIATION IN ITS RELATION TO EVOLUTION.

CHAPTER XI.

THE ACTION OF NATURAL SELECTION ON VARIATIONS.

CHAPTER XII.

ADAPTIVE VARIATIONS.

VARIATION IN ANIMALS AND PLANTS.

PART I.

THE FACTS OF VARIATION.

CHAPTER I.

THE MEASUREMENT OF VARIATION.

Variation studied from the mathematical standpoint—Variation of birds diagrammatically represented—Distribution of crab measurements—Normal curve of error—Its relation to binomial curve—Variations in animals and plants subject to Law of Frequency of Error—Measurement of variation in terms of Probable Error, Arithmetic Mean Error, and Error, of Mean Square—Examples of asymmetrical series—Representation of these by a generalised mathematical expression.

If a number of individuals of any species be compared, it will be found that they all show differences from each other either in size, shape, colour, relation of parts, or other characteristics; in fact, no two of them are exactly alike. Even if offspring be compared with their own parents, similar, though on the whole not such marked, differences will present themselves. These differences constitute what is known as *Variation*, and it is into the facts of this variation, and its importance as the corner stone of the whole fabric of Evolution, that we shall briefly inquire in the following pages.

In his "Origin of Species," Darwin clearly recognised the fundamental importance of the existence of variation, for without it there could evidently be no such thing as evolution. In his "Variation of Animals and Plants," also, he brought together an enormous mass of material concerning the facts of variation, though unfortunately this dealt almost exclusively with organisms in a condition of domestication. Still, there was sufficient evidence even then to show that wild animals and plants are also subject to variation, though Darwin probably did not fully recognise how considerable and universal this variation is. As to the causes of variation, Darwin did not hazard many conjectures. To do so would have been premature, and from actual lack of knowledge almost impossible. For many years after the publication of Darwin's work, the additions to our knowledge of the subject of variation were exceedingly small. Scientists seemed to rest content with the material he had collected, and to theorise on this alone, rather than to test their theories by a search after fresh facts and data. Within the last decade, however, the importance of the scientific study of variation has begun to be more thoroughly recognised, and has resulted in its being attacked with considerable vigour from several entirely different points of view. Investigations from the mathematical side have shown that many of the apparently disconnected facts of variation can be expressed with ease and lucidity by exact mathematical expressions, and that much material which has hitherto been regarded as quite outside all law was in reality amenable to treatment according to the well-known Laws of Chance. Again, investigations from the experimental side have suggested much concerning the causes of variations, both genetic and somatic. Still again, a fresh burst of activity in the collection of data regarding the actual facts of variation, more especially in respect of organisms found in a state of nature, has shown us how much in this branch of the subject there remains for us yet to learn.

Perhaps the keynote of most of the recent work on variation lies in the recognition of the fact that almost all the problems to be solved must be attacked from a numerical standpoint. It is no longer sufficient to say that such and such a kind of variation is frequently or occasionally found. It is necessary to know the exact amount of the variation, so far as it is measurable, and the exact proportion of cases in which it occurs. Only by obtaining data of this kind can we hope to ascertain with any certainty the probable degree of importance of any particular variation in the evolution of a species, and whether such evolution is actually taking place at the present day. No apology is therefore needed for the frequent introduction of figures into the study of questions of variation. Rather is this necessary if one should attempt to found theories and deduce conclusions from generalised statements and opinions, unsupported by such evidence. To say that any particular organ is very variable means but little, for so much depends upon the personal opinion of the observer as to what constitutes a great and what a slight variation. But supposing it be said that out of a large number of individuals half varied in size by ± 5 per cent. from the average of the whole, then there is afforded a numerical expression of the degree of variation, which can readily be compared with similar expressions concerning the variability of other parts of the same organism, and with those of quite distinct organisms.

Let us first of all, therefore, examine one or two simple series of measurements made on a group of individuals of a species, so as to get some idea of the actual differences exhibited by the varying characters, or, as they have been termed, the *variants*. Some of the most striking are those obtained by J. A. Allen,* concerning the variation in certain mammals and winter birds of East Florida. Of a species of squirrel (*Sciurus carolinensis*), for instance, 28 individuals were measured, and these measurements are reproduced to scale in the accompanying diagram. Here the animals are arranged in order according to the length of their body in inches, and the corresponding values for the head, tail, and forefoot are given on the same ordinates. By means of this diagram, the magnitude of each and all of the measurements made can be read off at a glance. The body was on an average 9.15 inches long, but the extreme values were 8.25 and 10.20 inches, or respectively 9.8 per cent. and 11.5 per cent. less and greater than the mean. The tail measurements were even more variable than this, the extremes varying from 6.75 to 8.75 inches, or by respectively 14.3 per cent. and 11 per cent. from the mean. In the forefoot the range of variation was less, and in the head smaller still; but

* Bulletin Museum Comp. Zoöl., Harvard, 1871.

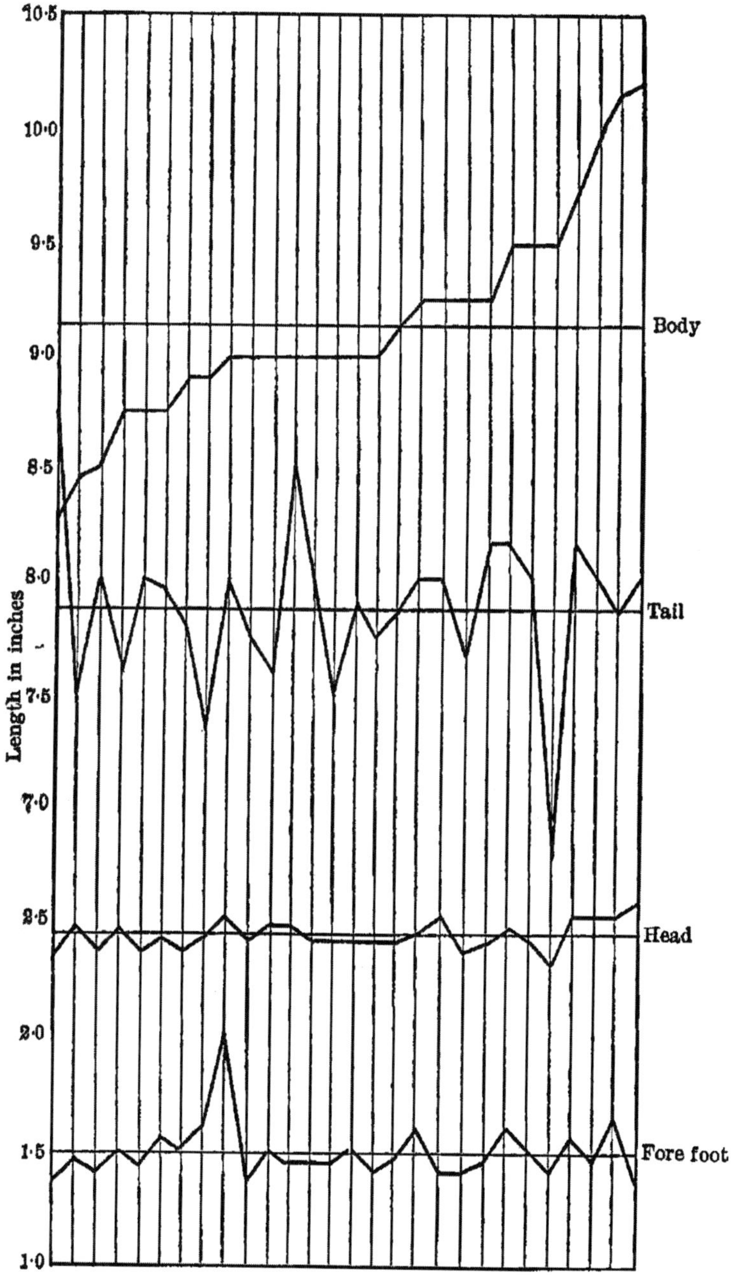

FIG. 1.—Variation of *Sciurus carolinensis.*

there was never any constancy, every animal varying in respect of each of the measurements made. This is a point of fundamental importance, which cannot be too thoroughly grasped. Every organism varies in respect of all its characters, whatever be their nature. The amount of this variation differs greatly, as these results well show, but it is always present in a greater or less degree. Another fact which this diagram brings out very clearly is the comparative independence of these measurements. Because the body of one animal is longer than another, it by no means necessarily follows that the head or tail is longer also. A superficial glance at this diagram might, indeed, lead one to suppose that the various parts of the body were absolutely independent of each other. But this we know not to be the case. Between most parts and organs there is a greater or less degree of correlation, so that, on an average, animals with a longer body may have a longer head and longer tail than animals with a shorter body. A careful examination of the diagram will show that, on the whole, though with numerous exceptions, the curves for head, tail, and foot do slope very slightly upwards from left to right, though nothing like as much as the curve for body lengths. Some degree of correlation is therefore present, though it is only slight. We know that frequently it may be very great indeed, as for instance between the two fore limbs or two hind limbs of a quadruped, and very considerable between a fore and a hind limb; but into this question we must not enter now.

Almost innumerable diagrams of a similar nature to the above might be given, but this is scarcely necessary. All that they would demonstrate would be the fact that variation of a similar nature—though of a varying degree—is present in all organisms, to whatever class of the Animal or Vegetable Kingdom they belong. Should more evidence of this kind be desired, the reader is referred to Wallace's book on "Darwinism" (Chapter III). Here an admirable series of diagrams is given, illustrating the variation in several species of lizards, birds, and mammals. The diagram given above is modelled on the plan adopted by Wallace, and still earlier by Galton, as the one best adapted for bringing before the eye the facts of individual variability.

In the above diagram the measurements of only 28 different individuals are given, and hence we are not able to gather much as to the distribution of the different measurements about their means. Supposing that instead of tens, fifties or hundreds of the animals had been measured, what should we expect to find? Would there or would there not be just as many animals with a very long or very short body length, as with a moderately long or moderately short one, or as with a nearly average one? Such a question as this is also best answered by reproducing the measurements diagrammatically, though in this case they must be arranged on a different system. In the accompanying diagram, Fig. 2, 65 measurements of the wing of *Sterna hirundo*, recorded in the above-mentioned paper of Allen, are plotted out. Here each dot represents one measurement, all the measurements between 10.46 and 10.55 inches being placed over the number 10.5, and so on. The mean of all the measurements is 10.49 inches, and we see in the diagram that the most frequently occurring measurement is one of 10.5 inches. Wing lengths smaller or greater than the mean occur less and less frequently, in rough proportion to their degree of deviation from it, so that finally, beyond the extreme deviations of 9.6 and 11.7 inches, no measurements were observed at all.

The number of observations here plotted out is obviously much too small to yield at all a regular series, but it is quite sufficient to show that the

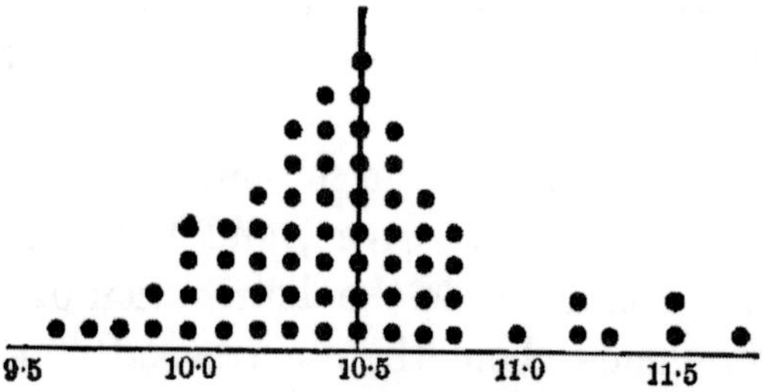

FIG. 2.—Wing of *Sterna hirundo.*

measurements are by no means evenly distributed through the whole range of their variation. There is a most conspicuous collection of them, or heaping up, in the region of the mean measurement. Supposing the number of observations were increased, then one would expect as a general rule to get a more and more even series; in fact, to get a fairly accurate idea as to the kind of series obtainable, supposing an infinite number of observations were made. In Fig. 3 is plotted out a curve representing the distribution of 1923 measurements made by Warren * on a certain dimension, viz., the carapace breadth of the crab *Por-*

* Proc. Roy. Soc., vol. lx. p. 225.

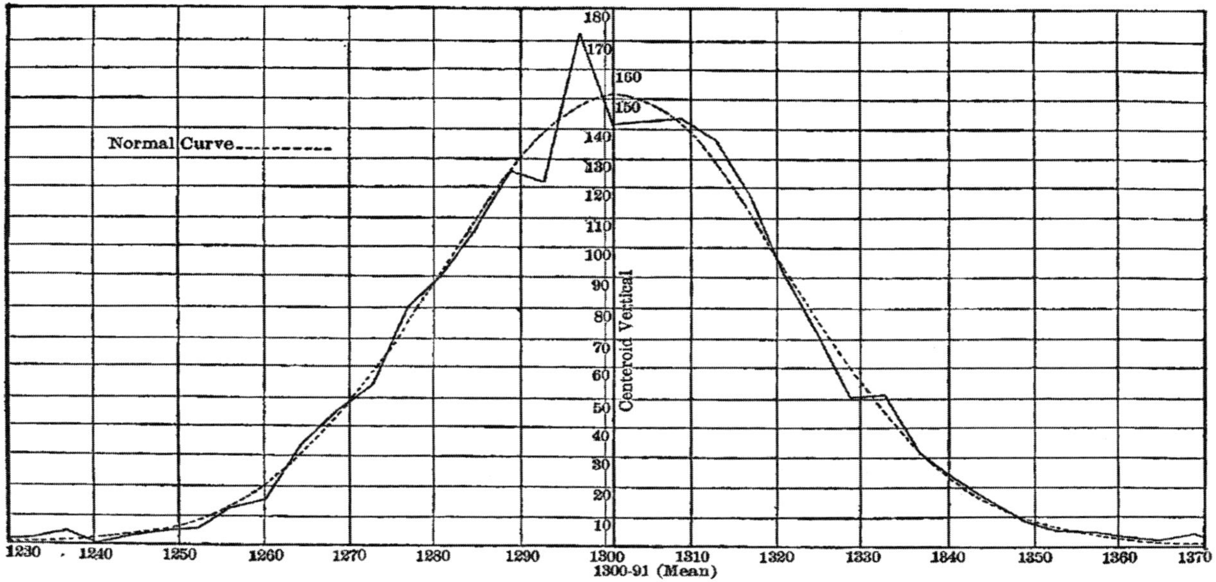

FIG. 3.—Distribution of carapace breadths of *Portunus depurator*.

tunus depurator. In order to get rid as far as possible of the factor of size, and obtain a measure of the variability apart from this, each measurement was calculated as a fraction on that of the carapace length of the crab taken as 1000. The numbers on the abscissa line therefore represent 1230, 1240, etc., thousandths of the total length. The figures on the central ordinate represent the numbers of individuals of each particular dimension. For instance, one may gather that 16 individuals had a post-spinous length of 1260, 172 of them one of 1297, and so on.

If this curve be compared with the general contour of the previous figure, it will be seen at a glance that there is a much more regular rise and fall, especially in regard to the extreme measurements. In fact, it does not differ very greatly from the dotted line curve upon which it is superposed, and supposing the number of observations had been greater, one would expect the approximation to be still closer; supposing it had been infinitely great, one would expect the two curves to be identical. Now this dotted line is a probability curve, or a diagrammatic representation of the Law of Frequency of Error, of which the mathematical expression* was first deduced by Gauss at the beginning of the last century. It would be out of place to attempt to reproduce its mathematical proof here, but perhaps a concrete instance may help to bring home to the non-mathematical reader the fact that variability does obey

* This expression is $y = ke^{-h^2 x^2}$, or taking k and h each as unity, $y = \frac{1}{e^{x^2}}$, where e is the base of Naperian logarithms, and y an ordinate erected from any point on the abscissa, distant x from the middle ordinate.

the laws of chance. Supposing a group of developing organisms be taken, of which the growth can be affected in a favourable or an unfavourable manner by their surroundings. Let us suppose that there are twenty different agencies, each of which would produce an equal, favourable effect on growth, and twenty which would produce just as great an effect in the opposite direction. Suppose also that each organism is subjected to only half of these forty different agencies; then it would follow, according to the laws of chance, that a larger number of the organisms would be acted upon by 10 favourable and 10 unfavourable agencies, than by any other combination; *i. e.*, they would, on our hypothesis, remain absolutely unaffected in their growth. A somewhat smaller number would be acted upon by 11 favourable and 9 unfavourable agencies, or on the whole, would have their growth slightly increased. A still smaller proportion would be acted on by 12 favourable and 8 unfavourable agencies, or would have their growth rather more increased. Finally the number of organisms acted on by 20 favourable and 0 unfavourable agencies would be extraordinarily small, but in this case the effect on growth would be extremely large. Similar relationships, only in the reverse direction, would of course be found in those cases in which the number of unfavourable agencies exceeded the number of favourable. If desired, the proportional numbers of organisms acted on by all the different combinations of agencies may be readily determined by expanding the binomial $(\frac{1}{2} + \frac{1}{2})^{20}$. It is found, for instance, that for each single time the organisms are acted on by the whole 20 favourable agencies, they are acted on 190 times by 18 favourable and 2 unfavourable, 15,504 times by 15 favourable and 5 unfavourable, and no less than 184,756 times by 10 favourable and 10 unfavourable. Let us consider that the organisms acted on by 20 favourable and 0 unfavourable agencies have their size increased by 20 per cent., those acted on by 15 favourable and 5 unfavourable by 15 — 5 = 10 per cent., and so on. If now these percentage increments and decrements be plotted out at equal distances on a base line, and ordinates corresponding to the theoretical frequencies erected from each, then by joining these ordinates we shall obtain a curve which is practically identical in form with the dotted line curve given in Fig. 3; *i. e.*, with the probability curve of the law of frequency of error. Thus, by a simple arithmetical method, we can obtain a series approximating more and more closely to the probability curve, the greater the number of times the expression $(\frac{1}{2} + \frac{1}{2})$ is expanded. Expanded 20 times, the average error is less than .5 per cent., and for a greater number of times it becomes rapidly smaller and smaller.

The deviations in the dimensions of organisms are thus distributed about their mean in a symmetrical manner, in accordance with the law of frequency of error. This is true not of one or two characteristics of an organism, but probably, in the majority of cases, of nearly all of them. The dependence of variation on the Laws of Probability was first demonstrated by Quetelet * in the case of height and chest measurements of soldiers. These he showed to group themselves in accordance with the ordinates of a binomial curve.

* "Lettres sur la théorie des probabilités," Brussels, 1846.

Subsequently * he proved that a similar relationship was true not only for the height, weight, strength, longevity, and other physical qualities of man, but also for his intellectual and moral qualities, such as age at marriage, age of criminals, and so on. He considered also that these laws extend to the whole Animal and Vegetable Kingdoms, though he did not give proofs of this hypothesis.

In confirmation and extension of Quetelet's results, the observations of Mr. Francis Galton † may be quoted. These were made at the Anthropometric Laboratory of the International Health Exhibition of 1884, upon from 489 to 1788 men and women. It was found that the variations in height, span of arms, weight, breathing capacity, strength of pull, strength of squeeze, swiftness of blow, and keenness of sight all conformed in their distribution to the Law of Error. With regard to the lower animals, Professor Weldon ‡ has made measurements on the carapace, post-spinous portion of carapace, length of the sixth abdominal tergum, and length of telson, in the case of two to five local races of shrimps, and obtained a similar result. He has also § made no less than eleven different series of measurements on 999 female crabs (*Carcinus mænas*) obtained from Plymouth Sound, and a similar number on 999 specimens obtained from the Bay of Naples. Twenty series of frequencies of deviation from the average were thereby obtained, and were found in every

* "Anthropométrie," p. 257, 1870.

† "Natural Inheritance," p. 201.

‡ Proc. Roy. Soc., xlvii. p. 445, 1889, and Proc. Roy. Soc., li. p. 2, 1892.

§ Proc. Roy. Soc., liv. p. 318, 1893.

case but one to conform to that required by the Law of Error. The single exception to the general rule will be referred to in the next chapter. Again, H. Thompson * made twenty-two different measurements on 1000 adult prawns, and found the variations in every case but one to correspond more or less accurately with the law. E. Warren † made seven different measurements on 2300 male crabs (*Portunus depurator*), obtained from Plymouth, and found that the variations very nearly corresponded to the law. Duncker ‡ made eight series of determinations on the number of spines and rays in the fins of 1900 specimens of the fish *Acerina cernua*, and found that with one slight exception the variations obeyed the general law. Of the twelve series of measurements § made on 1120 specimens of the flounder (*Pleuronectes flesus*), however, only six were quite symmetrical and in accordance with the law. Finally the author ‖ made 9850 measurements on the plutei or larvæ of a sea-urchin, *Strongylocentrotus lividus*, and found that the variations in size corresponded very closely indeed with the law. The lengths of the arms of these plutei were calculated as percentages on the length of body, and were found in the case of the oral arm lengths to correspond closely with theory, but in the case of the anal arm lengths, there was some slight divergence.

With regard to the variation of plants, our accurate

* Proc. Roy. Soc., lv. p. 234, 1894.
† Proc. Roy. Soc., lx. p. 221.
‡ Biologischen Centralblatt. xvii. p. 785, 1897.
§ Zool. Anzeig., xxiii. p. 141.
‖ Phil. Trans. 1895, B. p. 613.

knowledge is derived chiefly from the work of Ludwig, De Vries, and Vöchting. The majority of variations hitherto examined have not been found to be at all accurately in accordance with the law of frequency of error, for reasons which will be referred to later. However, in the case of one or two local races of *Torilis anthriscus* (hedge parsley) examined by Ludwig *, the distribution of the frequencies of the numbers of branches in the main umbels more or less conforms, and the same is true for the numbers of ray florets in a pure race of *Chrysanthemum segetum* (corn marigold) examined by De Vries.† Again H. Vöchting ‡ has recently examined the anomalies occurring in 61,736 flowers of *Linaria spuria* (toadflax), obtained in different years and from different sources. He determined the proportions of the various forms of peloric flowers and anomalous zygomorphic forms, of flowers of varying structure and with various numbers of spurs, and came to the conclusion that their distribution followed the law of error. For instance, the numbers of flowers in each inflorescence showed the following variations:

Number of flowers,	2	3	4	5	6	7	8
Frequency,	1	6	283	61,060	221	9	1
Per cent.,	.0016	.0097	.459	99.153	.358	.014	.0016

Here we see that though more than 99 per cent. of the flowers exhibited the normal pentamerous form, yet the variations from this normal are very evenly distributed on either side of it. The distribution of the numbers in all the peloric flowers (*i. e.*, regular

* Bot. Centralb., vol. lxiv. p. 40.
† Arch. f. Entwickelungsmechanik, ii. p. 52, 1896.
‡ Jahrb. f. wiss. Bot., Bd. xxxi. p. 391, 1898.

flowers, instead of the normal irregular ones) was as follows:

Number of flowers,	2	3	4	5	6	7	8
Frequency,	1	2	43	810	52	2	1
Per cent.,	.109	.219	4.720	88.913	5.708	.219	.109

From these two series a very interesting relationship declares itself, which may for convenience be referred to here, though it properly comes under the heading of "correlated variations." Thus, as the following figures show, we find that the probability of occurrence of a peloric flower increases according to the amount of deviation of the number of flowers on a stalk from the normal pentamerous form, or that the less often a particular number of flowers occurs, the more frequently does it produce peloric flowers:

.132	per cent.	of the	5	flower	form	have	peloric flowers.
15.19	"	"	4	"	"	"	"
23.53	"	"	6	"	"	"	"
22.22	"	"	7	"	"	"	"
33.33	"	"	3	"	"	"	"
100.00	"	"	2 and 8	"	"	"	"

Of English observers, J. H. Pledge * has determined the variations in the numbers of petals, stamens, and carpels in 1000 specimens of *Ranunculus repens* (creeping crowfoot), the distributions of all but the numbers of petals agreeing fairly closely with the probability integral. For instance, the numbers of sepals varied thus:

Sepals,	3	4	5	6	7
Frequency,	1	20	959	18	2

We see, therefore, that in the majority of the characteristics of the various organisms investigated,

* Nat. Science, vol. x. p. 323, and vol. xii. p. 179, 1898.

especially those belonging to the Animal Kingdom, the variations are distributed about their mean in accordance with the Law of Error. It is scarcely necessary to point out, however, that the actual range of the variations is exceedingly variable, and that the general contour of the curves, supposing the results are expressed in that way, must be equally variable. The greater the variability of any characteristic, the more spread out, or flattened, must be the curve representing the frequencies of its deviations. If, therefore, results were invariably expressed in the form of curves, and if, by multiplying each series of measurements by some factor, the central ordinate were always brought to the same height, then it would follow that the variability of each characteristic would be accurately represented by the extent of spread of the curve. In order to obtain an index of the variability of any characteristic, we must accordingly adopt some convenient method of determining the degree of spread of its curve. One of the simplest of these methods, and one widely employed by English statisticians, is to determine the so-called *Probable Error*. The meaning of this term is best explained by reference to the accompanying diagram of a curve of frequency of error. The ordinate drawn through the middle of the curve is spoken of by Mr. Galton as the *Median*, and is denoted by the symbol M. In symmetrical curves it is identical with the ordinary arithmetic mean or average, and in this sense is called the *Centroid Vertical*. It is the middle value of the whole series of observations, which are symmetrically distributed on each side of it. That is to say, 50 per cent. of all the observations fall below it in magnitude, and 50 per cent. above it. The actual number of observations made is obviously represented by the area of the figure enclosed by the curve and the abscissa line, or the so-called "polygon of variation." The area to the left of the median corresponds to the half of the observations of less magnitude than the average, and

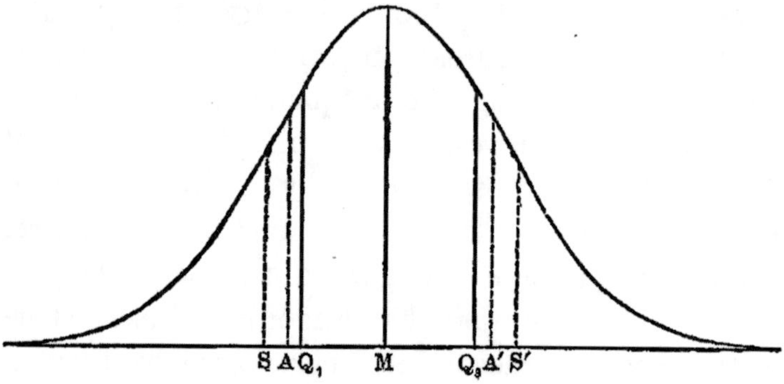

FIG. 4.—Normal Curve of Error.

that to the right, of those of greater magnitude. Now let two other ordinates, Q_1 and Q_3, be erected so as to divide each of these areas into equal halves. We now have four areas representing four numerically equal groups; *i. e.*, all the observations of small magnitude from 0 to 25 per cent. of the whole; those of greater magnitude, from 25 per cent. to 50 per cent. of the whole; those of greater magnitude than the average, representing 50 per cent. to 75 per cent., and finally those of greatest magnitude, representing the remaining 25 per cent. Half of all the observations therefore exceed the limits of these ordinates Q_1 and Q_3, and half of them fall between or within them; so the distance on the abscissa line from M to Q_1 or M to Q_3, is called the "Probable Error" of variation. In a perfectly normal curve, these values are equal in value and opposite in sign, but as no experimental result is perfect, they usually differ slightly in amount. A mean between the two is therefore taken, and this is denoted by the symbol Q.

For the practical determination of the probable error, however, it is quite unnecessary to plot out the results in the form of a curve. The method adopted is best illustrated by a concrete instance. In the accompanying table are given the results obtained by Mr. Galton * for the strength of pull, as of an archer with a bow, of 519 males, aged 23 to 26:

STRENGTH OF PULL.	NUMBER OF CASES OBSERVED.	PERCENTAGES.	
		NUMBER OF CASES OBSERVED.	SUMS FROM BEGINNING.
Under 50 lbs.	10	2	2
" 60 "	42	8	10
" 70 "	140	27	37
" 80 "	168	33	70
" 90 "	113	21	91
" 100 "	22	4	95
Above 100 "	24	5	100

Here we see that the numbers of actual cases in each group are given in the second column, and that they are calculated as percentages in the third column. They are summed from the beginning in the fourth column, and we thereby gather that whilst only 37 per cent. of all the men had a strength of pull under 70 lbs., 70 per cent. of them had one under 80 lbs. It can be calcu-

* "Natural Inheritance," p. 199.

lated, therefore, that 50 per cent. of them had a strength of pull under 74 lbs., or, in Mr. Galton's notation, the strength of pull at *Grade* 50°, was under 74 lbs. This, then, is the average strength of pull, or M, of the whole group. Fifty per cent. of the men pulled less than this amount, and 50 per cent. of them more. Similarly, also, one can calculate that 25 per cent. of the men would have a pull of less than 66 lbs., and 75 per cent. one of greater amount, whilst 75 per cent. would have one of less than 82 lbs., and 25 per cent. one of greater. That is to say, the strengths of pull at Grades 25° and 75° were respectively 66 and 82 lbs. The probable error of variation in pull, or Q_1, is therefore equal to $74 - 66 = 8$ lbs., and also to Q_3, or $82 - 74 = 8$ lbs., whilst the mean value which is always in practice adopted as the probable error, or Q, is $\frac{8+8}{2} = 8$ lbs. This probable error is 10.8 per cent. on the magnitude of the average strength of pull, and this value accurately represents the variability of this group of men in respect of this particular characteristic. Supposing another group were found to have a probable error of only 5.4 per cent. on the magnitude of the average, then one would be justified in saying that their variability, or range of variation, was only half as great; or if it had been 21.6 per cent., then twice as great.

This relative probable error is therefore a convenient index of variability of any characteristic. A few examples may be quoted in order to give an idea as to its range. From the anthropometric data obtained by Mr. Galton, it is calculated that the index was 2.50 per cent. for man's stature, and 2.52 per cent.

for woman's; 2.92 per cent. for the span of arms of both man and woman; but no less than 6.89 per cent. for man's weight, and 8.89 per cent. for woman's. That is to say, weight is more than twice as variable as the other two characteristics. In most of Weldon's shrimp and crab measurements the amount of variability was considerably smaller, but this was partly due to the fact that the element of size was largely excluded by first of all calculating all measurements as thousandths of the body and carapace lengths respectively. In 1000 shrimps from Plymouth, the total carapace length had a relative probable error of only 1.82 per cent., the post-spinous carapace length one of 1.97 per cent., the sixth abdominal tergum one of 1.93 per cent., and the telson one of 2.36 per cent. In 999 crabs obtained from Naples, the value was only 1.07 per cent. for the total breadth of carapace, and from 1 to 2 per cent. for several of the other measurements made, but in the carpopodite of the right chela it rose to 3.63 per cent., and in the proximal portion of the chela to no less than 5.77 per cent. These last were, however, quite exceptionally large degrees of variation. Still, in the sea-urchin larvæ measured by the author, the variability was found to be greater than in any of these instances recorded in the higher animals, it being 6.1 per cent. for the body length, 9.4 per cent. for the oral arm length, and 11.3 per cent. for the anal arm length.

We have seen that the degree of correspondence of the variations in any characteristic with the law of error can be determined by plotting out the results in the form of a curve, but this is clearly a somewhat laborious process. It is much simpler and more convenient to compare the experimental and theoretical values directly by a numerical method. This is done by extending the method of grades referred to above. In addition to determining the magnitude of the characteristics at grades 25°, 50°, and 75°, one determines it also at grades 5°, 10°, 20°, 30°, and so on, or determines the values having respectively 5, 10, 20, 30 per cent., etc., of all the measurements below them in magnitude, and 95, 90, 80, 70 per cent., etc., above them in magnitude. Let the median, or value at grade 50°, be now subtracted from the values at all the other grades, and the numbers so obtained be divided by the probable error, or $\frac{Q_1 + Q_3}{2}$. We then obtain a series of values at the various grades, in terms of the probable error taken as unity; so that, whatever had been the magnitude of the median, and of the probable error, the values are now directly comparable with the theoretical values calculated from the probability integral. These theoretical values are given in the first line of the subjoined table:

GRADE.	5°	10°	20°	25°	30°	40°	50°	60°	70°	75°	80°	90°	95°
THEORETICAL VALUES	2.44	1.90	1.25	1.00	.78	.38	.00	.38	.78	1.00	1.25	1.90	2.44
9443 anthropometric measurements, .	2.44	1.87	1.24	1.00	.77	.40	.00	.38	.75	.98	1.21	1.92	2.47
400 shrimp carapace length measurements	2.42	1.86	1.22	1.00	.79	.39	.00	.32	.71	1.00	1.28	2.10	2.63
9850 measurements of sea-urchin larvæ, .	2.51	1.92	1.25	1.01	.79	.38	.00	.37	.77	.99	1.24	1.90	2.46

Beneath them are given the means of the values obtained by Mr. Galton for 18 different series of measurements on men and women, the total number of observations made being 9443. In the individual series the deviations from the theoretical values were of course greater, but these differences almost completely neutralise each other in the general mean. Indeed the correspondence is extraordinarily close, considering the very mixed nature of the faculties measured, viz., three linear measurements, one of weight, one of capacity, two of strength, one of vision, and one of swiftness. The next series of values is that obtained by Professor Weldon for 400 shrimps. It is given to show that in the case of a comparatively small number of observations, the correspondence between fact and theory may be very close indeed. Finally, in the bottom line of the table are given the values obtained by the author for 9850 measurements on the body length of sea-urchin larvæ. Here the correspondence is closer even than in the anthropometric measurements, the average difference being only .014, as against .0175.

In order to express the variability of a characteristic, we are by no means limited to the method of determining the probable error. A much older method is that of the arithmetic mean error, or average deviation. This value consists of the mean of all the deviations, both positive and negative, from the general mean. For instance, to determine the arithmetic mean error of the following series of 16—

7, 8, 8, 9, 9, 9, 10, 10, 10, 10, 11, 11, 11, 12, 12, 13.

figures, one calculates the general mean, viz., 10, and determines their deviations from it. These are

3, 2, 2, 1, 1, 1, 0, 0, 0, 0, 1, 1, 1, 2, 2, 3.

Added together these equal 20, so that the arithmetic mean error is $\frac{20}{16} = 1.25$. In practice, it is sometimes simpler to separate all the numbers into two groups, one containing all the values greater than the general mean, and the other all those less than the mean. Then the arithmetic mean error is half the difference between the mean of each group. When, as in the present instance, several of the values are identical with the mean, half of them must be put in each group. The mean of one group is now $\frac{70}{8} = 8.75$, and of the other group $\frac{90}{8} = 11.25$, and the arithmetic mean error is

$$\frac{11.25 - 8.75}{2} = 1.25.$$

This method of estimating variability has frequently been employed in recent times, especially in America. Thus Minot * used it for comparing the variability of guinea-pigs at various periods of their growth. Brewster † used it for calculating the amount of variation in a number of head, face, and limb measurements which were made by Weisbach ‡ on individuals of 23 different races of men. In the general mean are included the measurements of 195 individuals, representing 20 different races. The following are some of the mean values for the arithmetical mean error, calculated as percentages on the mean size:

Nose length,	9.49	per cent.	Head length,	2.44	per cent.
" breadth,	7.57	"	" breadth,	2.78	"
" height,	15.2	"	Upper arm length,	6.50	"
Forehead height,	10.4	"	Forearm length,	3.85	"
Underjaw length,	4.81	"	Upper leg length,	5.00	"
Mouth breadth,	5.18	"	Lower "	5.04	"
			Foot length,	5.92	"

* J. Physiol., xii. p. 138, 1891.

† Proc. Amer. Acad. Arts Sci., xxxii. p. 268, 1897.

‡ Zeitschrift f. Ethnologie, Bd. ix. Supplement, 1878.

Here we see that most of the face measurements are far more variable than most of the head and limb measurements; that of the nose height, for instance, being six times as great as that of the head length. The high value which is universally accorded to facial proportions as a means of personal identification thus receives its numerical justification.

On comparing the variability of the measurements in the individuals of eight different races, it was found to be more or less the same in each case. If the nose of a Jew is a very variable organ, so is that of a Slav, a Magyar, or a Chinaman.

Davenport and Bullard * used the method of arithmetic mean error in the 4000 enumerations which they made of the Mullerian glands in the forelegs of swine. These glands vary in number from 0 to 10, the average being 3.53. The arithmetic mean error was 1.41 in male swine, and 1.38 in female swine, or the variability was 2.5 per cent. greater in the one case than the other. Again Garstang † has used it to estimate the variability of various local races of the mackerel.

There is still another method of estimating variability, which is more accurate than either of the two mentioned, but which until recently has not been used so frequently as they were, because of the labour of applying it. This is the method of Mean or Least Squares. One determines the deviations from the average in the same way as for the arithmetic mean error, but then squares each of them, takes the sum of these squares,

* Proc. Amer. Acad. Arts Sci., xxxii. p. 87, 1896.
† Jour. Marine Biol. Asso., vol. v. p. 235, 1898.

divides by the number of observations, and takes the square root of the quotient. Thus:

$$\varepsilon \text{ (or } \sigma) = \sqrt{\frac{\Sigma (v^2)}{n}}$$

Where n = number of observations, and v = a deviation from the average. For instance, to determine the variability of the following series, representing the frequencies of the numbers of veins in 26 leaves collected from different parts of a beech tree,* we find the

Number of veins,	15	16	17	18	19	20
Frequencies,	1	4	7	9	4	1

mean (17.5), determine the deviations from it in each direction, and square them. Then the variability will be represented by the square root of the following expression:

$$\frac{(2.5)^2 \times 1 + (1.5)^2 \times 4 + (.5)^2 \times 7 + (.5)^2 \times 9 + (1.5)^2 \times 4 + (2.5)^2 \times 1}{26}$$

i. e., by 1.15.

This index of variability, or "Error of Mean Square," is termed by Professor Pearson the "Standard Deviation," or σ, and its percentage ratio on the mean the "Coefficient of Variation." It has been made use of by Warren in the crab measurements already referred to, and also in a very elaborate research † on the variability of the skeleton of the Naquada race, a people that existed in Egypt about 3500 B. C. It has also been employed by Weldon, whilst Pearson almost invariably adopts it. Duncker ‡ has expressed his

* *Vide* K. Pearson's "Grammar of Science," p. 382.
† Phil. Trans. 1898, B. p. 135.
‡ Biol. Centralblatt, xvii. p. 785, 1897.

results on the variability of the fin rays of certain fishes in terms both of the mean error and the error of mean square. Again, from data obtained by Petersen, Bumpus, Weldon, and himself, Duncker * has calculated the error of mean square, and obtained the following values for the number of fin rays in certain fishes:

	DORSAL M	FIN ε	ANAL M	FIN ε
Pleuronectes flesus, Baltic,			39.46	1.4838
" " North Sea,			41.56	1.7739
" " Plymouth,	61.72	2.3895	43.61	1.6026
" *americanus*,	65.06	2.4467	48.62	1.8188
Rhombus maximus,	62.98	2.2533	45.86	1.6792

And the following for the number of rostral teeth:

	DORSAL M.	FIN ε	ANAL M.	FIN ε
Palæmonetes varians,	4.3137	.8627	1.6948	.4799
" *vulgaris*,	8.2819	.8145	2.9781	.4477

These results show that though the average values of a character may differ considerably even in the local races of the same species, yet the indices of variability may remain fairly constant, not only in these, but also in different species, and perhaps even in different genera and families. Thus the two species of *Palæmonetes* vary by respectively 92.0 per cent. and 75.7 per cent. in the number of rostral teeth in their dorsal and anal fins, but by only 5.9 per cent. and 7.2 per cent. in their indices of variability. Arguing from these data, Duncker † concludes that one has no right to accept the "coefficient of variation" of an organ as the

* Nat. Science, xv. p. 328, 1899.
† Amer. Nat., xxxiv. p. 621, 1900.

absolute measure of its variability, as Verschaeffelt,* Brewster,† and others have done. He thinks that the indices of variability alone may be of morphological significance, for in this case, at least, they are obviously independent of the mean values of the characters.

How far Duncker's view is to be accepted can only be determined by further enquiry. Doubtless it will be found to hold good occasionally, but I think that the great weight of evidence at present available, especially as regards measurements of size and shape, rather than those of numbers of organs, is in favour of the alternative hypothesis.

The three indices of variability above referred to are by no means numerically equivalent. They bear the following relations to each other:

Probable error,	1.000	Corresponding	grades,	25°.0, 75°.0
Arithmetic mean error,	1.183	"	"	21°.2, 78°.8
Error of mean square,	1.483	"	"	16°.0, 84°.0

Thus the error of mean square is nearly half as large again as the probable error, and therefore includes a proportionately larger percentage of the deviations from the mean within its limits. On the frequency curve given a few pages back are drawn dotted line ordinates A, A′ and S, S′, which enclose areas of the variation polygon corresponding to these "mean error" and "error of mean square" indices of variability. The "probable error" index is in some ways the most convenient of the three, as it is the smallest, and includes within its limits just half of all the variants. As the error of mean square is held to be a more accurate

* Ber. d. deutsch. bot. Ges., xii. p. 350.

† *Loc. cit.*

method of estimating variability, however, the plan is sometimes adopted of determining this index, and then reducing it to terms of probable error by multiplying by .6745. Similarly an arithmetic mean error may be reduced to terms of probable error by multiplying by .8453.

It will have been noticed that in the series of measurements from time to time referred to, a few exceptions to the general law of distribution of variations were mentioned. In these cases the variations were not distributed evenly about the middle ordinate, but the curve of distribution was asymmetrical, or skew. Such series as these are by no means uncommon, especially in the case of plant statistics. For instance, De Vries * found that the number of petals in the buttercup varied between 5 and 10, the frequency of distribution being as follows:

Number of petals,	5	6	7	8	9	10	11
Frequency observed,	133	55	23	7	2	2	0
Theory, .	136.9	48.5	22.6	9.6	3.4	.8	.2

Here flowers with the smallest number of petals occur the most, and those with the largest number the least, frequently. The values marked "Theory" in this and the next series will be referred to later.

Again De Vries cultivated a variety of clover in which the axis is very frequently prolonged beyond the head of the flower, and bears from one to ten blossoms. The following were the frequencies of occurrence of flowers with none of these blossoms, or with various numbers of them:

* Ber. d. deutschen bot. Gesellschaft, xii. p. 203, 1894.

High blossoms,	0	1	2	3	4	5	6	7	8	9	10
Frequency obsd.	325	83	66	51	36	36	18	7	6	1	1
Theory,	303.2	106.1	70.0	49.3	35.2	24.9	17.1	11.0	6.3	2.8	.5

J. H. Pledge * observed the following frequencies in the numbers of petals in *Ranunculus repens:*

Number of petals,	4	5	6	7	8	9	10	11	12	13
Frequency,	8	706	145	72	38	15	7	7	1	1

Again E. T. Browne † found the following variations in the number of tentaculocysts in the ephyra and adult forms of the medusa *Aurelia aurita:*

Tentaculocysts,	4	5	6	7	8	9	10	11	12	13	14	15
Percentage in 1136 ephyræ,	0	.09	.5	3.0	79.1	6.7	5.4	3.1	1.4	.2	.09	0
Percentage in 3000 adult *Aurelia*,	.1	.1	.7	4.1	78.9	6.3	4.8	3.0	1.4	.4	.1	.1

In each case the normal eight tentaculocyst form comprised nearly four-fifths of the whole, but the great majority of the abnormal forms possessed more than eight tentaculocysts, only 3.6 to 5.0 per cent. of them having less.

Now the distribution of frequencies in these and somewhat similar asymmetric series obviously occurs according to some orderly plan, but can a mathematical expression be obtained to represent them? This had been found impossible till within the last few years, when Professor Pearson ‡ took up the subject, and showed that such series, if composed of homogeneous material, could often be fitted most exactly with curves calculated in accordance with a single generalised

* Nat. Science, vol. xii. p. 179, 1898.

† Q. J. Microsc. Sci., vol. xxxvii. p. 245, 1895, and Biometrika, I. p. 90, 1901.

‡ Phil. Trans. 1895, A. p. 343.

mathematical expression. We saw a few pages back that an expansion of the binomial $(\frac{1}{2} + \frac{1}{2})$ for 20 or more times gave a series of values which differed very slightly in their frequencies from that required by the Law of Error. Supposing, now, a binomial in which the two terms are unequal is expanded, then obviously an asymmetrical series of values is obtained. For instance, instead of $(\frac{1}{2} + \frac{1}{2})$ let $(\frac{3}{4} + \frac{1}{4})$ or $(\frac{7}{8} + \frac{1}{8})$ be expanded, and series are obtained of which the diagrammatic representations are given in the two curves to the left of the accompanying figure. The symmetrical curve represents the expansion of $(\frac{1}{2} + \frac{1}{2})$, the expression being in each case expanded ten times. The areas enclosed between each of these curves and the base line, or the so-called polygons of variation, are obviously of exactly equal extent, in that the sum of the two terms expanded is in each case equal to unity.

It follows, therefore, that these asymmetrical series can be represented by the expansion of the expression $(p + q)^n$.* Supposing that n is infinitely large, then curves representing the expansion would stretch out to an unlimited extent in each direction, and though constantly approaching nearer and nearer to the abscissa, would never touch it. Supposing n is some finite number, as 20 or 40, then obviously the series is finite also, and its curve is limited in extent. If the two terms of the binomial are unequal, then the curve approaches the

* The algebraical expansion of this expression is:

$$(p+q)^n =$$

$$p^n + np^{n-1}q + \frac{n(n-1)}{\lfloor 2}p^{n-2}q^2 + \frac{n(n-1)(n-2)}{\lfloor 3}p^{n-3}q^3 + \dots$$

$$+ nq^{n-1}p + q^n.$$

abscissa much more rapidly on one side than on the other, and so, for practical purposes, by taking various values for p, q, and n, we can represent series of the

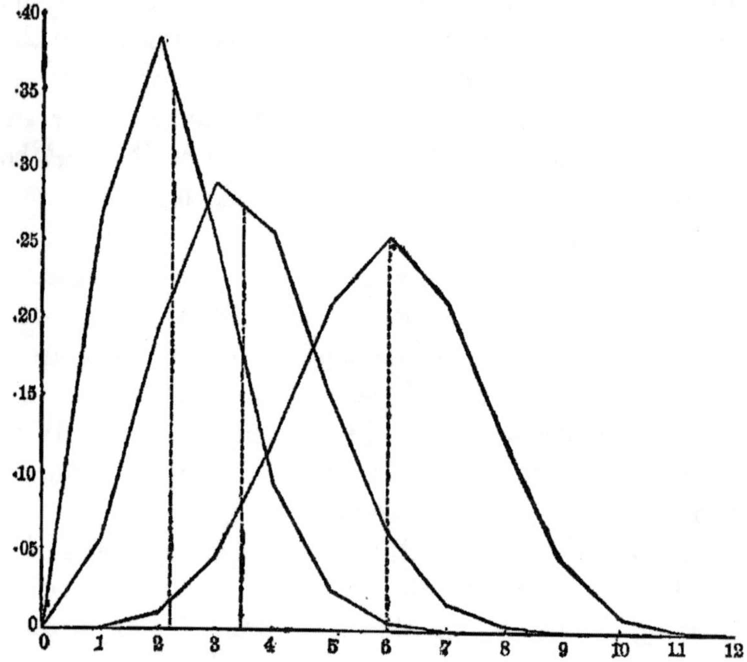

FIG. 5.—Types of binomial curves. (After Duncker.)

following five types by means of the above generalised expression:

I. Asymmetrical curves limited on both sides.
II. Symmetrical " " " "
III. Asymmetrical " " " one side, unlimited on the other.
IV. Asymmetrical curves, unlimited on both sides.
V. Symmetrical " " " "

The normal curve of error belongs to this last type. Pearson has also pointed out that the abnormal frequency curves which cannot be represented by a point-binomial may be the resultant of two or more normal curves, which differ in the position of their axes, or their areas, or their degree of spread, or in all three of these respects.*

To return to the curves in Fig. 5, we see that the centroid vertical of the symmetrical curve corresponds to the summit of the curve, or is identical with the maximum ordinate or *mode*, as it is sometimes called. In the asymmetrical curves, however, this is not the case, but the more asymmetrical the curve, the greater is the distance between the two. The ratio between this distance and the index of variability adopted (such as the error of mean square), gives a convenient "index of asymmetry" of the curve. It is to be noted also that in asymmetrical curves the median, or middle value of the whole series, such that 50 per cent. of the values are below it in magnitude and 50 per cent. above it, no longer coincides with the arithmetic mean. It lies somewhere between the centroid vertical and the maximum ordinate.

As to the practical application of this method of fitting series of variation frequencies with curves, Professor Pearson gives numerous instances in the above cited memoir. Fig. 6 will serve to afford some idea as to the types of frequency curves actually met with in practical statistics. Type α represents the above-mentioned series of frequencies which De Vries obtained for the petals of buttercups, and high blossoms of clover. It also represents infantile mortality statistics. Type β represents the relation of scarlet fever and diphtheria mortality to age; type γ that of scarlet fever and

* Proc. Roy. Soc., liv. p. 329.

typhus fever cases to age; type δ that of typhoid fever cases to age, and also senile mortality statistics. Finally, type ϵ represents various slight degrees of skewness which are frequently found to occur even in anthro-

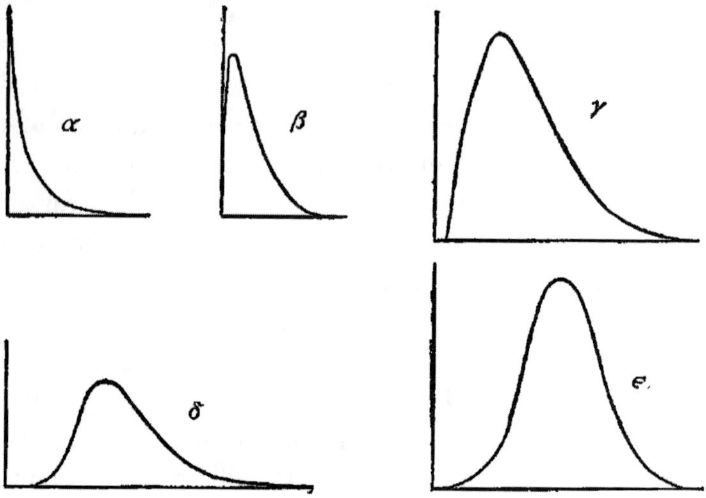

FIG. 6.—Types of Skew curves.

pometric and other series which had previously been thought to be quite symmetrical. Most of the series of deviation frequencies obtained by Warren * for various crab measurements were found by him to be better fitted by skew curves than by absolutely symmetrical ones.

Again, of the twelve series of measurements made by Duncker † on the Flounder (*Pleuronectes flesus*), the six which showed regular variations (number of rays in dorsal, anal, and pectoral fins) were found to give very

* Proc. Roy. Soc., lx. p. 221.

† Wissenschaftliche Meeresuntersuchungen aus der biologischen Anstalt auf Helgoland, Bd. iii. p. 339, 1900.

slightly asymmetrical curves of variation, the variability of the bilateral homologous measurements being always, with one exception, slightly higher on the blind than on the eye side of the fish. Still the departure of some of the curves from the normal Gaussian curve was only very slight. The degree of difference between the actual frequencies obtained in any series of measurements and the theoretical frequencies calculated from the type of curve found to show the closest agreement with them, is best represented by determining the percentage difference of each actual frequency from each theoretical frequency, and then taking an (arithmetical) average of the whole. This average percentage difference of theoretical and actual values may be represented by the sign Δ. In the case of four of the above series of measurements, the Δ was only 2.20 per cent. when the measurements were compared with a normal Gaussian curve, and 1.82 per cent. when compared with a slightly asymmetrical curve (Pearson's Type IV). The "fit" was therefore better with the asymmetrical curve, but only very slightly so. It is open to question, therefore, whether any practical object is gained by estimating exceedingly slight degrees of asymmetry in series of measurements. The labour of so doing is very considerable, and it may well be doubted whether it would not be more profitably employed in making more extended series of observations, and subjecting them to less rigid examination. Some recent observations of Miss Hefferan * are instructive in this connection. These were made upon the frequency of distribution of the numbers of teeth on the jaw of an annelid, *Nereis*

* Biol. Bulletin, vol. ii. p. 129, 1900.

limbata. Four hundred individuals were examined, and it was found that, as regards the distribution of the total number of teeth, the left total fell into a curve of Pearson's Type I, whilst the right total was of Type IV. However, by dropping out a *single* individual from the series, it was found that the curve was thrown from Type IV to Type I. As Miss Hefferan points out, this raises a serious question as to the biological importance of the distinction between Pearson's Type I and Type IV.*

* Should any further information regarding these asymmetrical curves be desired, the reader should consult Professor Pearson's memoir on the subject, or, if he is not a mathematician, then a recently published book by Davenport on "Statistical Methods,"* and also a paper by Duncker on "Die Methode der Variations-statistik"† may be referred to. Both of these are said to be written specially for biologists. I must mention my special indebtedness to Duncker's paper, which has been drawn upon freely in writing the last few pages of the present chapter.

* New York and London, 1899.

† Arch. f. Entwickelungsmechanik, Bd. viii. p. 112.

CHAPTER II.

DIMORPHISM AND DISCONTINUOUS VARIATION.

Dimorphism in the earwig and in the crab—How to distinguish between species and varieties, as instanced by dimorphism in certain fishes, and in a marsh plant—Polymorphism in plants—Series of Fibonacci—Discontinuous variation in animals as regards vertebræ, ribs, mammæ, teeth, digits, and other characters—Homœosis—De Vries' Theory of Mutation—Dimorphism may be due to internal causes, or the result of divergent evolution—Physiological Selection—Infertility between varieties.

WE have seen that the distribution of variations about their mean is in many cases quite symmetrical, whilst in other cases in which it is asymmetrical it still takes place according to some orderly arrangement, for which a mathematical expression can be found. There is still a third group of cases, however, in which the curve of distribution is, as a rule, very asymmetrical, but for which, even if symmetrical, no single general mathematical expression is obtainable. A study of such curves has taught us that the cause is frequently referable to the fact that our material is not homogeneous; that, in fact, we have a mixture of varying numbers of two or more groups of individuals differing in mean size and range of variation. For instance, Bateson * measured the length of the forceps of 583 specimens of the common earwig, *Forficula auricularia*, which had

* "Materials for the Study of Variation," p. 41.

been collected at random in one day in the Farne Islands. Only mature males with elytra fully developed were measured. The range of variation was from 2.5 to 9.0 mm., the various lengths occurring with a frequency indicated by the accompanying curve. Here

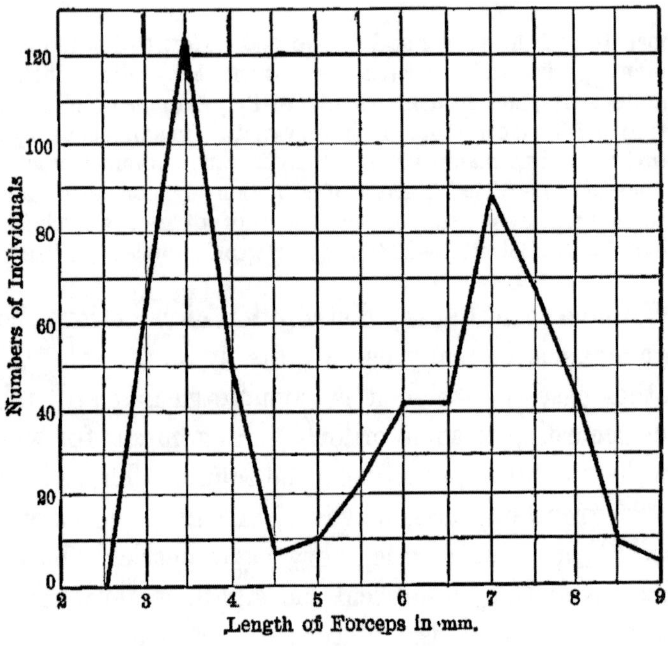

FIG. 7.—Distribution of various lengths of forceps in the male earwig.

we see that the commonest kinds had a length of either 3.5 mm., or 7 mm., or were of the forms shown in Fig. 8; whilst individuals of intermediate length occurred comparatively infrequently. This species was therefore most distinctly dimorphic in respect of the character measured, and it maintained this dimorphism in spite of the fact that the varying individuals were living in close communion with each other under the same stones.

In this instance there could be no doubt as to the dimorphism, just as in many of the instances previously quoted there could be practically no doubt as to the monomorphism, but obviously there must be intermediate stages in which the fusion is much closer. In these an irregular asymmetrical curve with only one summit may be obtained, instead of a distinct double

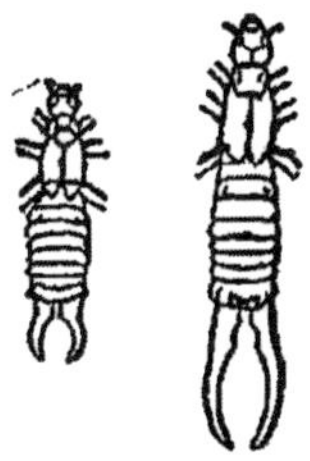

I II

Fig. 8.—I. High male, II. Low male, of Common Earwig.

humped curve. Weldon * obtained such a curve for the distribution of the frontal breadths of Naples specimens of the crab, *Carcinus mœnas.* This is reproduced in Fig. 9, the horizontal scale representing thousandths of the carapace length, and the vertical scale numbers of individuals. It seemed to Professor Weldon very probable that this asymmetrical curve was produced by the fusion of two races of individuals, clustered symmetrically about separate mean magnitudes, and Professor Pearson tested this supposition for him. Pearson calculated that by mixing 41.45 per cent. of individuals with a mean frontal breadth of 630.62 thou-

* Proc. Roy. Soc., liv. p. 324, 1893.

sandths, and a probable error of 12.06, with 58.55 per cent. of individuals of mean breadth 654.66 and probable error 8.41 (*i. e.*, groups of individuals represented by the two lower dotted line curves in the Figure), the upper dotted line curve would be obtained. It will be seen that this corresponds very closely with the ob-

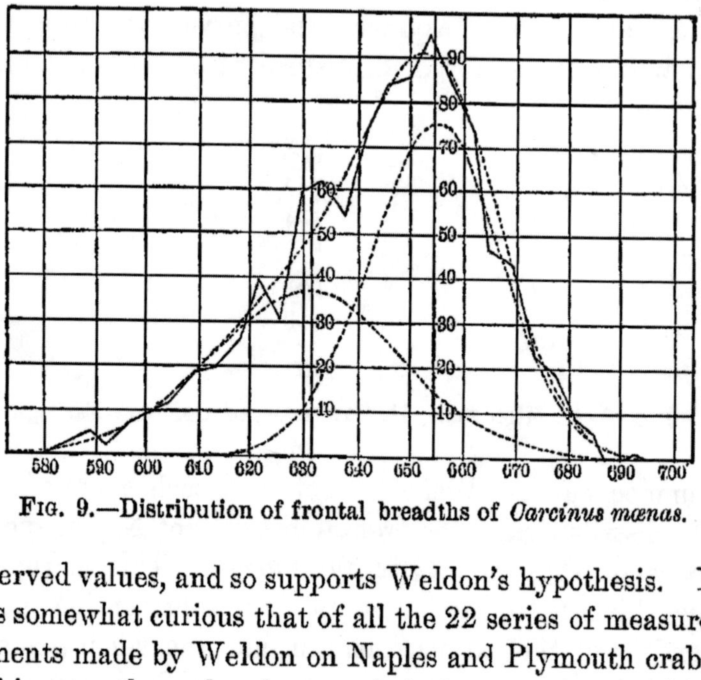

Fig. 9.—Distribution of frontal breadths of *Carcinus mœnas.*

served values, and so supports Weldon's hypothesis. It is somewhat curious that of all the 22 series of measurements made by Weldon on Naples and Plymouth crabs, this was the only characteristic in respect of which dimorphism was exhibited. As an explanation of it, Giard * has suggested that one of the two groups owed its altered frontal breadth to the presence of an internal parasite, *Portunion mænadis.* Thus he measured five specimens of *C. mænas* infested by this parasite, and found that their mean relative frontal breadth very

* Comptes Rendus, cxviii. p. 870, 1894.

nearly corresponded to the lower mean value of Weldon's crabs (viz., 630.32 as against 630.62). Giard thinks that the dimorphism in the length of the forceps of the earwig observed by Bateson can be similarly explained, for the short individuals appear to be infested with Gregarines, and the longer ones not. He does not wish to insist, however, that all dimorphism is the result of parasitic influence, but merely that it may be so in certain instances. It is obvious, indeed, that between two absolutely distinct varieties or species, and between pure monomorphic forms, all intermediate stages may exist. But how are these intermediate stages to be classified? When is one justified in assuming the existence of two distinct species, and when of only one species with an increased range of variation and perhaps a tendency to split up? To overcome this difficulty Davenport and Blankinship * have suggested that in order to decide in any given case whether we are dealing with two or more confluent species, or only with varieties, the following procedure should be adopted: First of all one should determine the most distinctive character of the members of the group, and after making a series of measurements in respect of this character, plot out a curve showing the relative frequency of occurrence of each measurement. Supposing that in this way a double humped curve is obtained, then the degree of isolation of the constituent races is estimated by measuring the depth of the depression between the two humps, from the level of the maximum of the lower hump. This depth may be expressed as a percentage on the length of the maximum ordinate, or

* Science, N. S. vol. vii. p. 685, 1898.

mode, of the lower hump. The value so obtained is termed by Davenport and Blankinship the "Index of Isolation." They suggest that if this index be over 50 per cent., then one should agree to look upon the two groups as distinct species; if under 50 per cent., then only as varieties. As an example, they adduce a case of two doubtful species of fishes, *Leuciscus balteatus*, and *L. hydrophlox*, which differ in the number of rays in the anal fin. On plotting out the frequencies of oc-

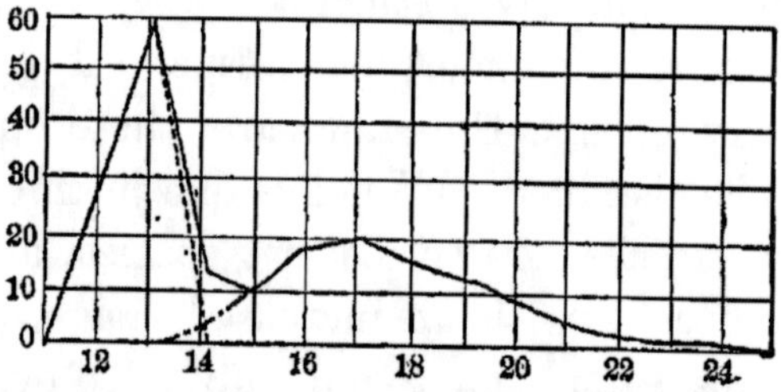

FIG. 10.—Distribution of fin rays in *Leuciscus*.

currence of the various numbers of rays in 194 individuals, the curve given in Fig. 10 was obtained. In this case the index of isolation is exactly 50 per cent., hence we are just at the limit of species and varieties.

The importance of determining which is the most distinctive character, before drawing any conclusions from the indices of isolation found, is well shown by another case adduced by these authors. It concerns the marsh plant *Typha*, which is found in the eastern United States. Seven characters regarded as probably specific were measured in about 250 specimens, which had been collected at distances of about one metre apart across the swamps in which it occurred. The variation curves obtained in the case of the stem height, diameter of stem taken at half the height, and the width of the largest leaf at its widest part, are reproduced in Figs. 11, 12, and 13. Here we see that the stem height shows no differentiation, the curve being more or less symmetrical. The mid-stem diameter shows a slight second hump, but this is obviously insufficient to indicate the presence of two species. The leaf width,

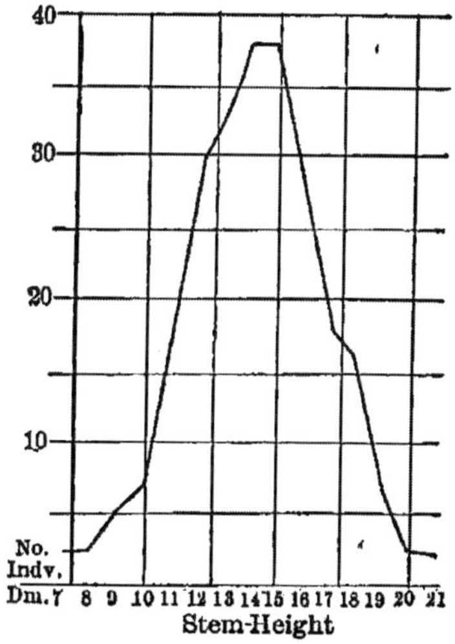

FIG. 11.—Distribution of Stem-Heights in *Typha*.

however, shows a marked differentiation, the index of isolation being 75 per cent. Of the other characters measured, the diameter of the stem at its base, the diameter of the pistillate spike, and the interval between the staminate and pistillate spikes, had indices of isolation of respectively 79, 89, and 83 per cent., or showed even greater differentiation than the leaf width. The curve for the pistillate spike length was, however, symmetrical. Thus this plant shows distinct differentiation into two species in respect of four out of the seven characters measured, and so is obviously, in the authors' opinion, to be regarded as composed of two more or less confluent species.

This "precise criterion of species" suggested by Davenport and Blankinship has much to recommend it, but probably it would generally be considered that an index of isolation of only 50 per cent. is too small a difference to merit specific distinction. Perhaps it would

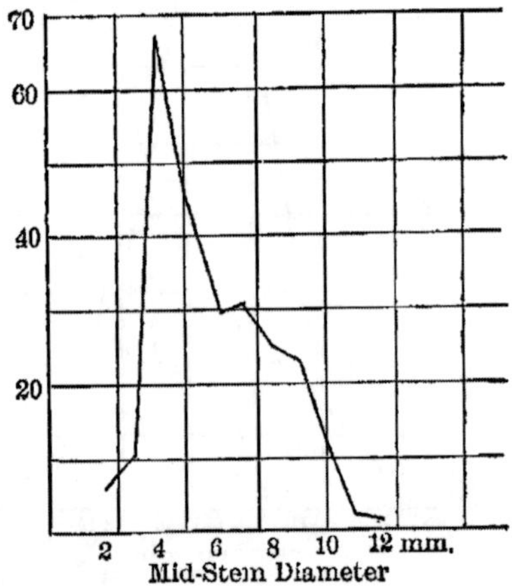

FIG. 12.—Distribution of Mid-Stem Diameters in *Typha*.

be better, therefore, to increase it to 90 or 95 per cent. In any case, it must, I think, be admitted that a slight degree of confluency, or overlapping of the curves of variation, ought not to compel one to assume the existence of only a single species, though this is the view which has been generally held in the past. On the other hand, the fact of a group of organisms showing absolutely discontinuous variation in respect of some apparently unimportant characteristic ought not to compel one to regard it as composed of two distinct species. It is, of course, often impossible to tell whether any given characteristic is important or not, and hence we must recognise that a really precise and univer-

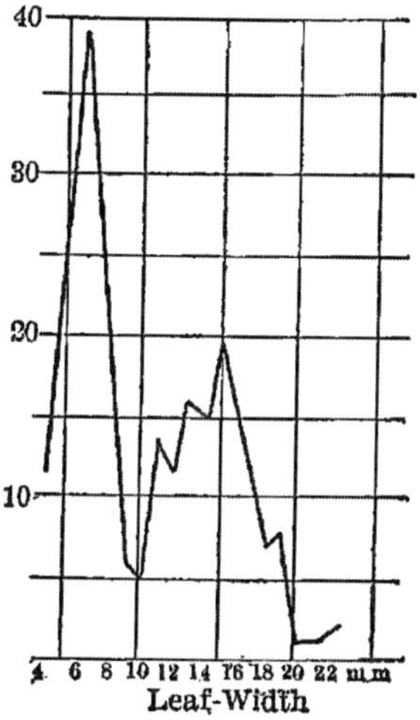

FIG. 13.—Distribution of Leaf-Widths in *Typha*.

sally applicable definition of a species is, and always must be, unattainable.

It is probable that variation series in the Vegetable Kingdom often give double humped or multiple humped curves, even if the material examined is as homogeneous as it is possible to obtain it. Possibly, if only individuals of the same stock were examined, they would be found to give single humped curves, but if material collected from different parts of the same district, or even of the same field, is to be regarded as composed of so many local races or sub-varieties, then the determination of the variations in many plant species would become an almost hopeless task. The degree to which local races may vary is well shown by some of Ludwig's determinations. For instance,* four groups of specimens of *Torilis anthriscus*, obtained from various spots near Schmalkalden, had the following numbers of branches of the main umbels:

	3	4	5	6	7	8	9	10	11	12	13	Total
Group I.			1	11	18	**45**	28	20	5	4	1	133
" II.				1	5	8	13	**25**	12	5	2	71
" III.				5	7	9	8	**12**	8	1	1	51
" IV.	7	60	**213**	152	46	18	3	1				500

In the first group, the 8 branch form occurred the most frequently, the variations from this number being distributed more or less evenly around it. In the next group, the 10 branch form was the commonest; in the third group the 8 and 10 branch forms were both common, whilst in the last group of all, collected in a wood at Wolfsberg, near to Schmalkalden, a quite distinct race having five branches presented itself.

In determining the variations of a plant, therefore, it is probably best to obtain a very large amount of material, from various sources, and submit this to examination. Though the curve thereby obtained may be very composite, yet at least it will indicate something as to the range of variation of the flower in many of its local races, and also what number of branches or other parts occur on the whole most frequently. For instance, Ludwig † has had made enumerations of the number of ray florets in 17,000 specimens of the Ox-eye Daisy, *Chrysanthemum leucanthemum*. The material

* Bot. Centralb., vol. lxiv. p. 40.

† Bot. Centralb., vol. lxiv. p. 1, 1895.

was collected from various sources, between the years 1890-95, and was examined by various people. The number of florets varied from 7 to 43, the following being the frequency of their occurrence:

Florets.	Frequency.	Florets.	Frequency.	Florets.	Frequency.	Florets.	Frequency.
7	2	16	479	25	602	34	346
8	9	17	525	26	614	35	186
9	13	18	625	27	375	36	64
10	36	19	856	28	377	37	28
11	65	20	1568	29	294	38	16
12	148	21	3650	30	196	39	16
13	427	22	1790	31	183	40	14
14	383	23	1147	32	187	41	0
15	455	24	812	33	307	42	3
						43	2

Those results are plotted out in the form of a curve in Fig. 14. Here we see that the 21 floret form occurs by far the most frequently, but that there are also secondary smaller maxima or humps on the curve for 13, 26, and 34 floret forms. This curve thus gives one a good idea both as to the range of variation of the number of florets in this flower, and also as to the most frequently occurring forms. Enumerations of small numbers of specimens of local races showed the 13 floret form to be the commonest form in one case, and the 34 floret form in another, but these results obviously fail to give a true idea of the variation of the plant.

Similar enumerations of the florets of various other species of the *Compositæ* showed the following to be the most frequently occurring numbers of ray florets, the absolute maximum being in each case indicated by thick type:*

Chrysanthemum leucanthemum,				13	**21**	34	26
" inodorum,				13	**21**		
" segetum, .				**13**	21		
Anthemis arvensis, . .		5	**8**	13			
" Cotula, . .			8	**13**			
" tinctoria, . .					**21**		
Achillea ptarmica, . .			**8**	13			10
Senecio nemorensis, . .	3	**5**					
" Fuchsii, . .	3	**5**					

Taking the observations as a whole, we see that the most frequently occurring numbers are the following:

3 5 8 10 13 21 26 34

In various species of the *Umbelliferæ* the following are the most frequently occurring numbers of petals:

3 5 (10 15 20 25) 8 13 21 34

That is to say, in each case the numbers follow the so-called series of Fibonacci, viz., 1, 2, 3, 5, 8, 13, 21, 34, etc., in which each number is the sum of the two before it; or else they follow some multiples of these numbers. Ludwig states that this relationship is by no means limited to the two orders of plants mentioned, but that it extends to other members of the Vegetable Kingdom, and probably also to the Animal Kingdom.

Perhaps the most striking result that Ludwig † obtained was for the number of petals of one of the Primroses, *Primula officinalis*. This varied from 1 to 22 in the sample of 1170 flowers examined. These flowers were all obtained from a single meadow (near Wieda), and were therefore as homogeneous as it was possible to

* Bot. Centralb., vol. lxiv. p. 100.

† Ber. d. deutsch. bot. Gesell., xiv. p. 204, 1896.

obtain them. The frequency of occurrence of the various forms is indicated in Fig. 15, given below. Here it will be seen that the curve has five most dis-

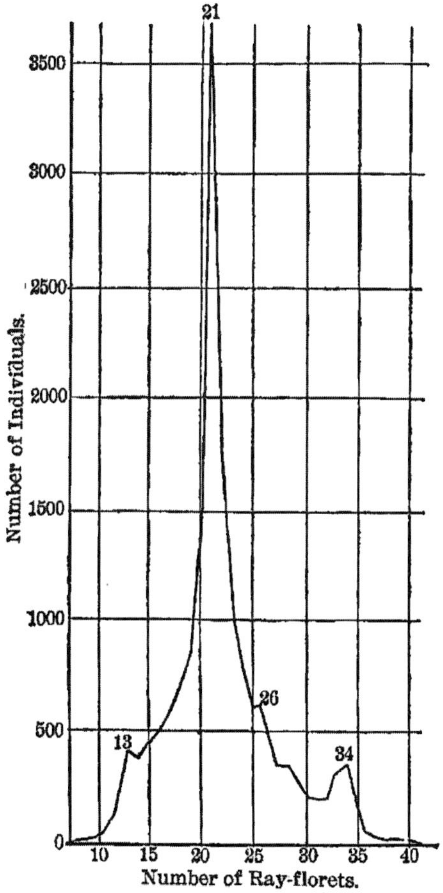

FIG. 14.—Distribution of Ray-florets in the Ox-eye daisy.

tinct maxima, corresponding to 3, 5, 8, 10, and 13 petal forms.

That these multi-humped curves are actually due to a mingling of two or more local races is proved by an interesting experiment of De Vries.* He sowed the mixed seed of *Chrysanthemum segetum* obtained from twenty different gardens. The topmost flowers of the

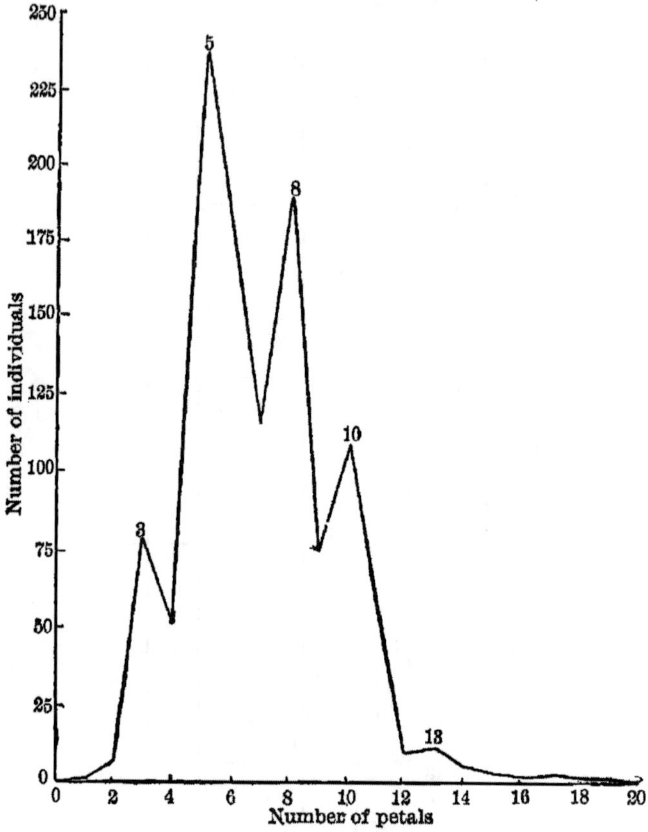

Fig. 15.—Distribution of Petals in *Primula officinalis*.

chief stem of each of the 97 healthy plants obtained were examined, and were found to contain the following numbers of ray florets:

Ray florets,	12	13	14	15	16	17	18	19	20	21	22
Frequencies,	1	14	13	4	6	9	7	10	12	20	1

*Arch. f. Entwickelungsmechanik, Bd. ii. p. 52, 1896.

Thus there were obviously two forms present, a 13 ray form and a 21 ray form. The seeds from the 12 and 13 ray forms were collected and sown next year, the flowers obtained therefrom having the following numbers of florets:

Ray florets,	8	9	10	11	12	13	14	15	16	17	18	19	20	21
Frequencies,	2	1	0	7	13	94	25	7	7	1	2	0	3	0

That is to say, all trace of the 21 ray form had been eliminated, and a nearly pure 13 ray form obtained. That this was so was proved by sowing the seed of some of the 12 rayed plants obtained on this occasion in the following year. It was then found that the frequencies of occurrence of flowers with various numbers of rays remained practically unchanged.

But how do these cases of what Bateson has termed *discontinuous variation* arise? In one or two of the instances quoted we saw that the two humps of the curve of variation scarcely overlapped at all. In the case of *Primula* they were all of them sharply defined, but there was still a good deal of fusion, whilst in the ray florets of the Ox-eye Daisy the fusion was greater still. Finally, in the frontal breadths of Naples crabs the fusion was complete, and the existence of dimorphism was shown only by the asymmetry of the curve. It would be possible to multiply instances of such curves as these, in which the fusion ran through all stages of completeness and incompleteness, but those quoted are quite sufficient for our purpose. They suffice to show that all stages of fusion *may* be met with, and so incline one to the opinion that the later stages, in which the two or more humps of the curve overlap little if at all, are but more advanced stages of those curves in which the fusion is nearly complete. In fact they seem to indicate that if only the ancestry of such varying organisms could be traced backwards continuously, it would be found that at no period was there any sudden change from continuity to discontinuity; that a condition of absolute dimorphism, or formation of two new species, was merely evolved by very gradual and almost imperceptible steps from the original pure monomorphism. This is, I believe, the opinion held by the majority of naturalists at the present day as to the origin of by far the larger number of cases of dimorphism, but dissentient voices have not been entirely wanting. Thus Galton * is of the opinion that the aberrant or discontinuous variations generally known as sports may be of considerable significance in evolution. Because evolution *may* proceed by minute steps, he considers that it does not by any means follow that it *must* so proceed. Again, within recent years the orthodox view has been ably combated by Bateson † in his book on Variation. In this work he has collected a very large number of instances of discontinuous variation, or variations in respect of certain organs or parts, which have suddenly arisen in a complete and perfect state, without, as a rule, the occurrence of any intermediate stages. If, therefore, argues Bateson, such instances of discontinuous variation undoubtedly occur, is it not possible that the Discontinuity of Species which is so striking a fact amongst living organisms is a consequence and expression of this discontinuity of variation? Thus the view

* "Natural Inheritance," p. 32, 1889.

† "Materials for the Study of Variation," London, 1894.

hitherto generally held, since Darwin first gave expression to it, is that almost all variations are very slight, and form a continuous series. It is only by their very slow accumulation, therefore, under the action of Natural Selection and other agencies, that species as we know them have been evolved. This view of Bateson's is so striking and important that it behoves us to examine it at some little length. It will enable us to obtain a clearer idea of Bateson's views if we indicate his system of classification. Thus he points out that variations are divisible into two classes, *substantive* and *meristic*. Substantive variations are variations occurring in the actual constitution or substance of the parts themselves. Meristic variations, on the other hand, are those which relate to the *number* of parts in organisms. For instance, the flower of the Narcissus is commonly divided into six parts, but through meristic variation it may be divided into seven parts, or only four. Nevertheless, there is in such a case no perceptible change in the tissue or substance of which the parts are made up. On the other hand many Narcissi, *N. corbularia*, for example, are known in two colours, one a dark yellow, and the other a sulphur yellow, though the number of parts and pattern of the flowers are identical. This is, therefore, an example of a substantive variation.

Bateson considers that there can be no doubt that these two classes of variation are essentially distinct from each other. It is obvious that all cases of meristic variation are also cases of discontinuous variation, whilst cases of substantive variation are much more frequently continuous than discontinuous. It is to be noted that these discontinuous meristic variations are not only large, but they are complete and perfect. But after all one would scarcely expect anything else. Between a six petal and a seven petal flower it is scarcely possible to imagine such a thing as a really intermediate stage. Even if one found a flower with six normal petals and a seventh abnormally small one, or five normal ones and a sixth in process of dividing into two, one would scarcely be justified in regarding it as an intermediate form, for the flower would no longer be symmetrical.

Perhaps the most interesting part of Bateson's work lies in the cases which he has collected of what he terms *Homœosis*. By this he means those variations which consist in the assumption by one member of a meristic series, of the form and characters proper to other members of the series. For instance, Kraatz has described a saw-fly, *Cimbex axillaris*, having the peripheral parts of the left antenna developed as a foot, the right antenna being normal. Kriechbaumer has described a nearly similar condition in a Humble-bee, *Bombus variabilis*. Bateson has himself described a crab, *Cancer pagurus*, having the right third maxillipede developed as a chela. Milne-Edwards has described another crab, *Palinurus penicillatus*, in which the left eye bore an antenna-like flagellum several centimetres in length, growing up from the surface of the eye. The eye stalk appears to have been of normal shape, but reduced in size. Other instances somewhat similar to these are adduced, but it is unnecessary to quote them here. It may be mentioned, however, that most interesting examples of this form of variation have been recently obtained by Herbst.* He found in no less than ten different species of Crustacea, belonging to four different families, that if an eye stalk were totally extirpated, there always grew up in its place a heteromorphic new structure, like an antennula, which bore olfactory hairs. If, however, only the eye were removed, and the stalk together with the ganglion left, instead of the antennula there arose the beginnings of a new eye. A similar result to this was obtained on extirpation of either stalk or eye in *Porcellana platycheles*, presumably because in this species the stalk contains no ganglion.

The chief contents of Bateson's book may be very briefly summarised, in order that the reader may gather some idea as to the kind of evidence on which Bateson founds his argument. A considerable body of evidence is given concerning variations in the numbers of vertebræ and ribs, the most important conclusions being that the variations are considerable, especially in some types such as *Simia satyrus*, the Bradypodidæ, and *Bombinator igneus*, and that imperfect vertebræ are very rare. Turning to Invertebrates, it is shown that among Oligochæta and Hirudinea, certain forms, *e. g.*, *Perionyx excavatus* and *Pachydrilus sphagnetorum*, have great variability, whilst others, such as the common earthworm, rarely vary. Both forward and backward Homœosis may occur, forms which normally have the male pores on the 15th segment having them on the 16th, or on the 13th. Returning to Vertebrates, evidence is next adduced concerning cervical fistulæ and auricles, and supernumerary mammæ. Variations in

* Arch. f. Entwickelungsmechanik, Bd. ix. p. 215, 1899.

teeth are dealt with in very great detail, and conclusions are drawn as to the comparative frequency of dental variation in various animals. The animals showing the greatest frequency of extra teeth are domestic dogs, Anthropoid apes, and the Phocidæ. It is especially noticeable that the variability of domestic animals in respect of teeth is not markedly in excess of that seen in wild forms. Thus, though supernumerary teeth are more common in domestic dogs and cats than in wild Canidæ and Felidæ, they are not more so than in Anthropoid apes and in the Phocidæ. With respect to the question of symmetry, the evidence shows that dental variation may be symmetrical on the two sides, but that much more frequently it is not so. Other evidence is given concerning the division of teeth, the presence and absence of first premolars and last molars, the least size of particular teeth, and other subjects.

Variations in the number of digits are treated more fully than any other subject discussed, though the evidence adduced is stated to bear rather on morphological conceptions than in any direct manner on the problem of Species. It is found that the frequency of digital variation is immensely greater in some classes of species than in others. Thus the horse shows many recorded cases, but the ass none at all. Variation is common in the cat, pig, fowl, and pheasant, but rare in the dog, sheep, and in most birds. In the cat, ox, horse, pig, and in man, the digital variation approaches to particular forms, and has in it something distinctive. Digital variation is sometimes symmetrical, but more often asymmetrical.

In other chapters of the book is found a considerable

mass of data concerning the repetition and division of appendages in insects and Crustacea, colour markings and colour variations in Lepidoptera, variations in the number of legs of different species of Peripatus, the occurrence of double monsters, and various other subjects, but to these it is unnecessary to refer here. Sufficient have been mentioned to indicate the general nature and scope of the evidence, so that we are enabled to enquire how far, if at all, it can warrant Bateson's hypothesis as to the origin and production of discontinuity in species. We see that most of the evidence concerns obvious abnormalities, generally in the direction of increase in the number of parts, which have arisen suddenly and apparently spontaneously. In practically no case has any new structure arisen, but only a repetition or misplacement of those already present, and so it is difficult to understand how really new structures and organs could have originated, even if it be admitted that such abnormalities are of very frequent occurrence, and that they could succeed in permanently establishing themselves. But first of all it is necessary to point out that the large majority of these abnormalities are extraordinarily rare, occurring perhaps not once in 100,000 or once in a million cases. What chance have they, then, of establishing themselves on a permanent footing? Bateson remarks, "An error more far-reaching and mischievous is the doctrine that a new variation must immediately be swamped," but he fails to adduce one tittle of evidence to prove that it is an error at all. This is simply because no such evidence exists. It is true that some animals are prepotent over others in procreating their characteristics, and their abnormalities if they possess them, but this prepotency is quite limited in its range. Darwin in his "Variation of Animals and Plants under Domestication" * mentions an instance of transmission of supernumerary digits through five generations, whilst in other cases they have reappeared after an interval of even three generations. "But," says Darwin, "we must not overestimate the force of inheritance. Dr. Struthers asserts that cases of non-inheritance and of the first appearance of additional digits in unaffected families are much more frequent than cases of inheritance."

Unless much stronger evidence than that hitherto advanced be obtained, it therefore follows that, according to the known laws of inheritance, suddenly occurring variations, unless artificially selected, must inevitably be swamped by intercrossing and disappear. Supposing, on the other hand, any such variation is artificially isolated, and bred in and in with its own offspring, then it may be possible to establish a distinct race, bearing in undiminished degree all the abnormal characteristics of the original variety. For instance, Darwin thus records the origin of the ancon sheep:† "In 1791 a ram-lamb was born in Massachusetts, having short crooked legs and a long back, like a turnspit dog. From this one lamb the *otter* or *ancon* semi-monstrous breed was raised; as these sheep could not leap over the fences, it was thought that they would be valuable. The sheep are remarkable for transmitting their character so truly that Colonel Humphreys never heard of 'but one questionable case' of an ancon ram and ewe

* Vol. i. p. 457, Ed. ii.
† *Ibid.*, vol. i. p. 104.

not producing ancon offspring." Again, Darwin says: * "It is certain that the ancon and the mauchamp breeds of sheep, and almost certain that the niata cattle, turnspit and pug-dogs, jumper and frizzled fowls, short-faced tumbler pigeons, hook-billed ducks, etc., suddenly appeared in nearly the same state as we now see them. So it has been with many cultivated plants." And then he adds, "The frequency of these cases is likely to lead to the false belief that natural species have often originated in the same abrupt manner. But we have no evidence of the appearance, or at least of the continued procreation, under nature, of abrupt modifications of structure."

We see therefore that, though Darwin brought forward much more powerful and convincing instances of discontinuous variation than those cited by Bateson, he held them to be quite inadequate to account in any way for the discontinuity observed in species.

Under the title of "*Die Mutationstheorie*," De Vries has recently promulgated views concerning the origin of species which are somewhat similar to those held by Bateson. The evidence he adduces in support of them is chiefly derived from observations of his own on flowering plants, and even if his theoretical views be entirely rejected, there can be no doubt as to the intrinsic interest and importance of the observations themselves. According to the theory of mutation, the qualities of organisms are built up of individual units sharply defined from each other. When, in the course of evolution, one species arises from another, it follows that the change takes place by a distinct step or jump,

* *Ibid.*, vol. ii. p. 409.

i. e., is discontinuous, and does not occur gradually. Such mutations may take place in all directions, but probably they only occur from time to time, due, perhaps, to the periodical action of fixed causes. They are distinct from the slight differences observed in local races and varieties, for these can be produced gradually by artificial selection and changed conditions of environment. Also Natural Selection can only lead to the formation of such local races, it being powerless to bring about true mutations. The variation which leads to the formation of new species, therefore, is essentially discontinuous, not continuous.

In order to obtain evidence in support of his theory, De Vries has cultivated over 100 different species of plants, but only one of them, *Œnothera Lamarckiana*, showed the desired mutations. This plant was originally brought to Europe from America, and kept under cultivation. It has since run wild, and De Vries obtained the stock of nine plants which formed his first generation from a field near Hilversum. Unfortunately, the true origin of the plant is obscure. In Britton and Brown's recently issued "Flora of the United States," no reference whatever is made to it as a wild species. Hence it is probably a garden variety of *Œnothera biennis* (Evening Primrose), and may be a hybrid plant, whilst the mutations obtained by De Vries may merely be partial or complete reversions to the original ancestors of the plant. To obtain these mutations, De Vries cultivated the plant through eight generations, and during this time, obtained over 50,000 specimens. Of these, 834 showed characters which sharply differentiated them from the normal *O. La-*

marckiana. De Vries classified them as follows: 350 *O. oblonga;* 229 *O. lata;* 158 *O. nanella;* 56 *O. albida;* 32 *O. rubrinervis;* 8 *O. scintillans;* and 1 *O. gigas.* Of these new "species," *oblonga, albida, rubrinervis, nanella,* and *gigas* remained absolutely constant in subsequent generations, when crossed among themselves, or self-fertilised, in the case of *O. gigas. O. scintillans* was not nearly so constant, the offspring yielding only about a third of the parent form, and the rest of them being *Lamarckiana, oblonga,* and *lata. O. albida* bred quite constant, but the plants were weak, and not very fertile.

De Vries looks upon his sports as true species, and not varieties, for he says that varieties differ from their parent species in only one or two characters, whilst species differ from their nearest allies in almost all their characters. Thus in comparison with the parent form, *O. Lamarckiana, gigas* was stronger and *albida* was weaker, both forms having broader and shorter leaves. The flowers of *gigas* were larger, those of *rubrinervis* darker yellow, those of *oblonga* and *scintillans* smaller, and those of *albida* paler. The cuticle of *albida* was rough. The bosses on the leaves of *lata* were increased, and on those of *scintillans* diminished. The formation of pollen was increased in *rubrinervis* and diminished in *scintillans.* The seeds of *gigas* were larger, and those of *scintillans* smaller; those of *rubrinervis* more abundant, and those of *lata* more scanty.

By artificial selection De Vries obtained in one instance * what he regards as a true mutation. This was in the case of *Linaria vulgaris* (yellow toadflax).

* "Die Mutationstheorie," p. 552.

Starting with plants which had one or two peloric flowers, he bred and selected them through several generations, and ultimately obtained some entirely peloric plants. These plants were most of them sterile, but a few yielded seeds, and ultimately De Vries obtained a peloric race, *Linaria vulgaris peloria*, only 10 per cent. of the seeds of which reverted to hemipeloric plants (*i. e.*, plants with some peloric and some non-peloric flowers). According to De Vries, this new race is a true mutation, because the pure peloric plants from which it was derived arose suddenly, and apparently capriciously, from hemipeloric parents. He also obtained new races of other plants by artificial selection extending through several generations, but these he regards only as varieties, and not true species, in that their formation was gradual. However, there seems to be no valid ground for sharply differentiating them in this manner. For instance, in the case of the five-leaved clover race (*Trifolium pratense quinquefolium*) obtained by him,* he started breeding with two naturally occurring clover plants which had four leaflets to their leaves, and in the case of one leaf, five leaflets. It is difficult to understand why these naturally occurring plants should not be regarded as true mutations, just as much as the peloric race above mentioned, or why, indeed, the naturally occurring hemipeloric plants from which the peloric race was obtained were not likewise true mutations.

In the case of the clover, the breeding was continued through several generations, the seed of only the few plants richest in four or more leaflets being preserved

* *Loc. cit.*, p. 437.

for sowing the following year. The proportion of four leaflet plants steadily increased, and in the fourth generation the most widely diverging plants had the following (percentage) proportions of leaves with from 3 to 7 leaflets.

Number of leaflets,	3	4	5	6	7	TOTAL NUMBER OF LEAVES COUNTED.
Normal plant,	17	16	**37**	14	16	172
Atavistic "	**75**	19	5	0	1	216
Extreme variation,	12	9	22	17	**40**	97

The plant considered as "normal" was obviously of a five leaflet type, the numbers of leaves with 3 and 4 leaflets, and those with 6 and 7 leaflets, being distributed symmetrically around it. A comparison of the numbers of leaflets in the "atavistic plant" with those in the "extreme variation" is interesting, as showing the range of variation possible in plants of the same stock.

As an instance of the formation of a variety, De Vries' experiments with *Chrysanthemum segetum grandiflorum* may be quoted.* Starting in 1896 with plants which had 21 ray florets occurring most frequently in their capitula, and none of which had more than 23 florets, he picked out each year the two or three plants richest in florets for breeding with, and sowed their seed the following year. In 1897 a single flower was obtained having 34 florets, but the 21 floret form was still the commonest. In 1898 one of the flowers had 48 florets, the commonest forms now having 26 or 34 florets. In 1899 one had 67 florets, the commonest forms having 26 or 33 to 35 florets, and in 1900

* *Loc. cit.*, p. 523.

one had 101 florets, the commonest form having 47 florets.

Let us now return to the cases of dimorphism mentioned at the beginning of the chapter. Instances were there adduced in which the dimorphism was slight, fairly marked, or so great that the two forms scarcely overlapped at all. To what may such dimorphism be due? Bateson points out that a dimorphic condition may have arisen from a previous monomorphic one, or it may always have been present since the character was first acquired. As already stated, the first view is the one which finds favour in the eyes of most biologists, but on the other hand there is a certain amount of evidence to show that the second view may hold good, at least in some cases. It is a well-known fact that when two breeds are crossed, their characters do not always blend, but are transmitted in an unmodified state to the offspring from one or from both parents. For instance, in breeding game fowls, if one crosses a black with a white game, birds of both breeds of the clearest colour are obtained. "Sir R. Heron crossed during many years white, black, brown, and fawn-coloured Angora rabbits, and never once got these colours mingled in the same animal, but often got all four colours in the same litter."* Again, Miss E. A. Saunders † has recently made observations on *Biscutella lævigata*, a cruciferous plant occurring as a perennial herb in the alpine and sub-alpine regions of middle and southern Europe. It was observed by Bateson that this species exhibits two distinct forms, which exist side by side, the one hairy

* "Animals and Plants," vol. ii. p. 70.

† Proc. Roy. Soc., vol. lxii. p. 11.

and the other smooth or glabrous. Intermediate forms were also found, but these were scarce. Some of the ripe seeds of these plants were obtained, and grown in England. A portion of the seedlings were derived from cross-fertilised seeds of known origin, and it was found that though there was to a certain extent a blending of parental characters as regards hairiness and smoothness in the offspring of plants of dissimilar types, giving rise to intermediate forms, yet this intermediate condition was found only quite exceptionally among full-grown individuals. It was much more common in the young plants, but as these grew older, their leaves became smooth, and hence almost all the plants were ultimately either hairy or glabrous; that is to say, they varied discontinuously.

Supposing a dimorphic condition is due to internal causes, or to the fact that it is the nature of the plant to vary in this way around two "positions of organic stability," as Galton has termed them, rather than around one such position, then it would seem almost impossible to get further to the root of such causes. Supposing, on the other hand, as is probably true in the majority of instances, this dimorphic condition has been derived from a previous monomorphic one, then we may hold a more reasonable hope of being able to elucidate the cause or causes of this evolution from one condition to another. The problem of the splitting up of species was recognised by Darwin to be one of immense importance, and he discussed it at some length in the "Origin of Species." * The chief cause of divergence of character he attributed to the circumstance that

*Ed. vi. p. 86.

"the more diversified the descendants from any one species become in structure, constitution, and habits, by so much the more will they be better enabled to seize on many and widely diversified places in the polity of nature, and so be enabled to increase in numbers." He also attached considerable importance to geographical isolation of a portion of a species, as an element in the modification of species through natural selection.

Though Darwin's principle of diversification of structure is doubtless a very true one, yet it does not in itself contain sufficient clue as to *why* a species should split up into two or more varieties. Thus, if by some means these actually arose, but both continued to inhabit the same area, it is difficult to understand why intercrossing should not rapidly reduce them to the single species from which they took their origin. It was to overcome this difficulty that Romanes suggested his theory of "Physiological Selection." * This theory is founded on the fact that individuals of a species, though fertile with some, may be perfectly sterile with other individuals, and this apparently independent of any differences of form, colour, or structure. Romanes thought that this incompatibility might run through a whole race or strain, and so a group of individuals of a species be in a physiological sense isolated from the rest, and therefore able to vary independently, without having their newly acquired characteristics swamped by intercrossing. As Wallace has very clearly shown, † this theory, in the form originally proposed by its au-

* Journ. Linn. Soc. (Zoöl.), vol. xix. p. 337, 1886. Also "Darwin and after Darwin," vol. iii. p. 41.

† "Darwinism," p. 181.

thor, cannot stand, at least for members of the Animal Kingdom in which there is no promiscuous union of the sexes. For instance, if 10 per cent. of the members of a species are thus physiologically isolated, so as to be fertile *inter se*, but sterile when crossed with any of the other members of the species, and if the interbreeding take place purely according to the laws of chance, then on an average only one-tenth of these 10 per cent. will happen to pair with individuals with which they are fertile, and the remaining nine-tenths will form absolutely sterile unions. Thus this physiologically isolated section will never be able to increase in numbers and establish itself. In the case of flowering plants which are fertilised by insects, each of which perhaps visits ten or more flowers in a journey, Fletcher Moulton has shown * that Wallace's objection does not hold, as the dimunition in fertility in such a case is practically negligeable. In his more recent discussion of the theory Romanes somewhat modified his views, and laid more stress on the fact that the mutual sterility may have been slight at first, and have been subject to a gradual development, it acting as a segregating cause in a degree proportional to its completeness. However, he makes no suggestion as to why and how such physiological incompatibility should arise, other than general ones such as the influence of food and climate, and spontaneous variability of the reproductive system.

It seems to me that the origin of this physiological barrier which so generally exists between species can be most readily accounted for by assuming that in some cases at least it is, as Romanes

* "Darwin and after Darwin," vol. iii. p. 165.

suggests, slowly evolved from an originally almost imperceptible degree of infertility, but that this takes place only simultaneously with the evolution of morphological character, in consequence of some form of isolation. Thus, suppose a number of individuals of a species become for a time separated from the remainder of the species by a geographical barrier, by migration, or some other cause of isolation, whereby they are enabled to vary independently of the general stock in response to changed conditions of life. Then as they gradually become more and more divergent from the parent stock in respect of morphological characters, it is highly probable that they may concurrently—perhaps from the direct action of the body tissues on the reproductive system—diverge also in respect of physiological characters. Should any of them now happen to meet and intercross with individuals of the parent stock, or even if they should occupy the same breeding area again, their newly acquired morphological characters would no longer be in danger of being swamped, for the simple reason that few or no hybrid offspring would result from such crossing. In the case of the higher animals, also, it is probable that individuals of different varieties or sub-species, once these are formed, instinctively tend to breed amongst themselves, and hence the chance of production of hybrid offspring is still further diminished. Thus Darwin records * that in Paraguay it is believed "that the native horses of the same colour and size prefer associating with each other, and that the horses which have been imported from Entre Rios and Banda Oriental

* *Ibid.*, vol. ii. p. 80.

into Paraguay likewise prefer associating together." Again, "It has been observed, in a district stocked with heavy Lincolnshire and light Norfolk sheep, that both kinds, though bred together, when turned out, in a short time separate to a sheep." Still again, with respect to fallow-deer, "Mr. Bennett states that the dark and pale coloured herds, which have long been kept together in the Forest of Dean, in High Meadow Woods, and in the New Forest, have never been known to mingle." Darwin adduces other similar instances, in the case of the dog, horse, sheep, rabbit, and pigeon; hence there can be little doubt of the genuineness of the phenomenon, even though it is not based on very exact observation.

Supposing that the above view is correct, it follows that between at least some varieties there must exist a greater or less degree of sterility. Of course it is not necessary that divergence of morphological character should *always* be accompanied by corresponding divergence of physiological character; but merely that this is sometimes the case. Upon this point Darwin has collected a considerable amount of evidence in his "Animals and Plants."* One or two of the cases there cited may be quoted here. Gärtner found that a variety of dwarf maize, bearing yellow seed, showed a considerably diminished fertility with a tall maize bearing red seed, though both varieties were perfectly fertile when crossed *inter se*. Again, in the genus *Verbascum*, numerous experiments were made by Gärtner with the white and yellow varieties of *V. lychnitis* and *V. blatteria*, when he found that crosses between similarly coloured flowers yielded more seed than those be-

*Vol. ii. p. 82.

tween dissimilarly coloured ones. These experiments have been repeated and extended by Scott with confirmatory results.

Still better evidence than that quoted by Darwin has been obtained by Jordan in a laborious research on various species of plants—annuals and perennials, bulbous and aquatic, trees and shrubs—extending over thirty years. Jordan found that when a Linnean species is indigenous to a country, and is of common occurrence, it is represented by more or less numerous and perfectly constant varieties, all growing in intimate association with one another. It was found that in many hundreds of cases these varieties, though they differed but slightly in morphological characters, came true to seed, but were always more or less infertile when crossed *inter se.**

With regard to members of the Animal Kingdom, there is very little evidence indeed. The following anthropological data may perhaps be held valid. From statistics collected in Prussia between 1875 and 1890, it was found that Protestants, Catholics, and Jews, when marrying among themselves, had, on an average, respectively 4.35, 5.24, and 4.21 children. When, however, the husband was a Jew and the wife a Protestant or Catholic, the numbers of children were only 1.58 and 1.38 respectively: and when the wife was a Jewess and the husband a Protestant or Catholic, only 1.78 and 1.66 respectively.† Whether this apparent partial sterility was due to differences of race or to social reasons, it is impossible to say. Again, Professor Broca ‡

* Quoted from Romanes, *ibid.*, p. 86.

† Quoted from Mayo Smith's "Statistics of Sociology," p. 115.

‡ "On the Phenomena of Hybridity in the Genus *Homo*," 1864.

has brought forward evidence that some races of man show diminished fertility together.

In order to obtain further evidence, the author * made numerous observations on the effects of crossing the colour varieties of the sea-urchins *Sphærechinus granularis* and *Strongylocentrotus lividus*. With the former organism the numbers of blastulæ and of larvæ produced on crossing dissimilar colour varieties were distinctly smaller than for similar varieties. In the most marked instance, the similar color varieties yielded on an average 98.5 per cent. of blastulæ, and 73 per cent. of larvæ, whilst the dissimilar yielded 68 per cent. of blastulæ and only 15.6 per cent. of larvæ. Moreover these latter larvæ were 4.5 per cent. smaller than the others. In the case of *Strongylocentrotus*, however, where the colour varieties are much less pronounced, there was very little difference of fertility.

There can be no doubt, therefore, that certain varieties show a greater or less degree of mutual infertility, though this is doubtless not nearly so marked, or of such frequent occurrence, as in the case of species. Whatever view be taken as to the cause of such infertility, and its relation to divergent evolution, these instances quoted have also an intrinsic value. They show that just as the deviations from the average in respect of morphological characters may form double humped curves, so the deviations in respect of physiological characters may show corresponding irregularities, though of course it is impossible to measure them exactly and construct their curves of variation.

* Phil. Trans. 1898, B. p. 511.

CHAPTER III.

CORRELATED VARIATIONS.

The measurement of correlation—Galton's function—Correlation between various organs in man, in local races of the shrimp, and in crabs—Comparison between primitive and civilised races of man—Correlation between morphological characters and the reproductive system—Genetic Selection in man—Especial fertility of type forms in certain plants—Evolution in the Peppered moth—Parallel variation—Importance of mathematical treatment of variation.

All parts of an organism are to a certain extent related to each other, so that when one part varies other parts vary simultaneously in a greater or less degree. That is to say, variations are correlated. The most marked and obvious correlation is that existing between homologous parts. The symmetry of the corresponding or homologous organs on the right and left sides of the body, which is present in most animals, represents a very close degree of correlation. But even in this case the correlation is not constant or complete. Thus the two arms and the two legs of a man resemble each other very closely indeed, but careful measurement shows that the resemblance is not absolute. Again, the arms, as a rule, vary in length more or less in the same proportion as the legs, but personal experience will probably recall instances to the contrary, in which the length of the limbs was quite disproportionate. Between the arms and the legs, therefore, the degree of correlation is obviously less close than between arm and arm, or leg and leg. Still again, personal experience teaches us that there is correlation between even the length of the face and that of the limbs. Tall men as a rule have longer faces than short men; or, a more striking instance, greyhounds have long heads and long legs on the one hand, as compared with bull-dogs with short heads and short legs on the other. Between the length of face and length of limb, however, it is clear that there is a less degree of correlation than between length of arm and of leg, and between certain other organs of the body the connection must be less intimate still. It follows, therefore, that between the various parts there may exist all degrees of correlation, stretching from an almost perfect degree of resemblance down to an absolute lack of it. We must also recognise the existence of negative correlation, in which the variation of one part in one direction is accompanied by a greater or less degree of variation of another part in the opposite direction. Here again we may experience all degrees of negative correlation, just as of positive. Instances of negative correlation are much less frequent than those of positive, and the only one known to me in the case of man is that recently discovered by Professor Pearson.* It was found that between stature and head index the correlation is distinctly negative, or that brachycephalic or relatively broad-headed persons are slightly shorter than dolichocephalic or narrow-headed.

From what has been said it is clear that a bald statement that in such and such a case one part or organ is

* Proc. Roy. Soc., lxvi. p. 23, 1900.

correlated with another conveys no exact meaning. Such a statement must vary according to the notion of the observer as to what does and what does not constitute correlation. In order to obtain reliable and comparable data concerning the degree of correlation, it is necessary to obtain a mathematical expression for it, just as one was found to be necessary for indicating the range of a variation. The fundamental theorems of correlation were for the first time exhaustively discussed by Bravais * more than half a century ago, but a more convenient and improved method of obtaining an expression was first indicated by Galton, and he termed it the correlation constant, or *r*. It is now more generally known as "Galton's function."

The principle on which this constant is determined is best explained by a concrete instance, viz., one given by Galton in the original paper in which he explained his method.† Galton's data are anthropometric ones, obtained at his own laboratory, and consist of several measurements made on 350 males of 21 years and upwards. For instance, Galton found that the average relation between stature and cubit, or distance between the elbow of the bent arm and the tip of the middle finger, was as 100 to 37. In determining the correlation between these two measurements, however, it is obvious that it is not possible to compare the absolute amount of variation of the one with the absolute amount of the other, or even the proportionate amounts, but

* "Analyse mathématique sur les probabilités des erreurs de situation d'un point." Mémoires par divers Savans, T. ix., Paris, 1846, p. 255.

† Proc. Roy. Soc., vol. xlv. p. 135, 1888.

we must first transmute them into units dependent on their respective scales of variability. We shall thus cause a long or a short cubit and an equally long or short stature, as compared to the general run of cubits and statures, to be designated by identical scale values. The most convenient unit to employ is the value of the probable error of each group. The probable error of the cubit is .56 inch = 1.42 cm.; and of the stature, 1.75 inch = 4.44 cm. Therefore each of the measurements of the cubit must be transmuted into terms of a new scale, in which each unit = .56 inch, and each of the measurements of the stature into those in which each unit = 1.75 inch. After this has been done, we shall find that on an average each deviation in the stature of say 1 unit from the mean is not accompanied by a similar deviation of 1 unit in the cubit, but by only .8 of a unit. Conversely it is found that in a similar manner each deviation in the cubit of 1 unit from the mean is accompanied by only .8 of a unit of deviation in the stature. The degree of correlation, or r, between the one organ and the other, is therefore said to be .8. If the correlation had been perfect, then this r would have been equal to 1, and if it had been entirely wanting, then it would have been 0. Comparison with other data shows that a correlation of .8 is a high one, not often surpassed. The other correlation constants determined by Galton are the following:

	MEAN r.
Stature and head length,	.35
Stature and middle finger,	.7
Middle finger and cubit,	.85
Head length and head breadth,	.45
Stature and height of knee,	.9
Cubit and height of knee,	.8

Here we see that the maximum amount of correlation was observed between stature and height of knee, and the minimum between stature and head length. Even in this latter instance, however, the correlation was fairly marked. Thus a constant of .35 indicates that in men 1.75 inch, or 1 unit, above the mean stature, the length of head will on an average be $\frac{.35}{1.00} \times .19 = .0665$ inch above the mean, .19 inch being the probable error of variation of the head length. The height of knee, on the other hand, would on an average be no less than $\frac{.9}{1.0} \times .80 = .72$ inch greater. The various medians or middlemost values and probable errors found by Galton are as follows:

	MEDIAN.		PROBABLE ERROR.	
DIMENSION.	INCH.	CENTIM.	INCH.	CENTIM.
Head length,	7.62	19.35	.19	.48
Head breadth,	6.00	15.24	.18	.46
Stature,	67.20	170.69	1.75	4.44
Left middle finger,	4.54	11.53	.15	.38
Left cubit,	18.05	45.70	.56	1.42
Height of rt. knee,	20.50	52.00	.80	2.03

In order to determine the degree of correlation between any two organs, it is therefore necessary to adopt the following procedure. Sort out all the individuals into groups such as, for instance, in the case of stature, those varying from 64 to 65, 65 to 66, 66 to 67 inches, and so on, and then determine in each of these groups the mean of all the deviations from the average of the organ of which the correlation with stature is to be determined. Thus, in the group of individuals 64 to 65 inches high, the average difference of all the individual cubit measurements from the median of the cubit (18.05 inches), is about .6 inch. Let the deviation of each value for stature from its median (67.20 inches) be now divided by the probable error of variation of stature (*i. e.*, 1.75 inch), and each associated mean deviation of cubit be divided by *its* probable error (*i. e.*, .56 inch). Then, by dividing each of these terms for cubit by the corresponding term for stature, a series of values is obtained, each representing the amount of correlation between the various degrees of stature and the cubit. These values would be approximately equal in amount if a very large number of observations were made, but with only moderate numbers they vary very considerably. A mean of all of them may be called r, or the average degree of correlation between cubit in relation to stature. In a similar manner the individuals must be split up into groups in respect of cubit, and the associated deviation of stature determined. Another series of correlation values will be obtained, representing stature in relation to cubit, of which the mean may be called r_2. This value is found to be approximately equal to r_1, and the mean of r_1 and r_2 is called r, or the correlation constant.

The degree to which these individual correlation values vary is best shown by means of a diagram. The one given in Fig. 16 is taken from Professor Weldon's paper on correlated variations in *Crangon vulgaris*,* and represents the correlation between the post-spinous carapace length and the total carapace length in Plymouth shrimps.

The mean value of r found was .81. In this diagram, the deviations of the organ whose value is fixed

* Proc. Roy. Soc., li. p. 2, 1892.

are measured along the ordinates, they varying on an average between the extremes of + 3.19 and — 3.01, whilst the mean deviations of the associated organ are

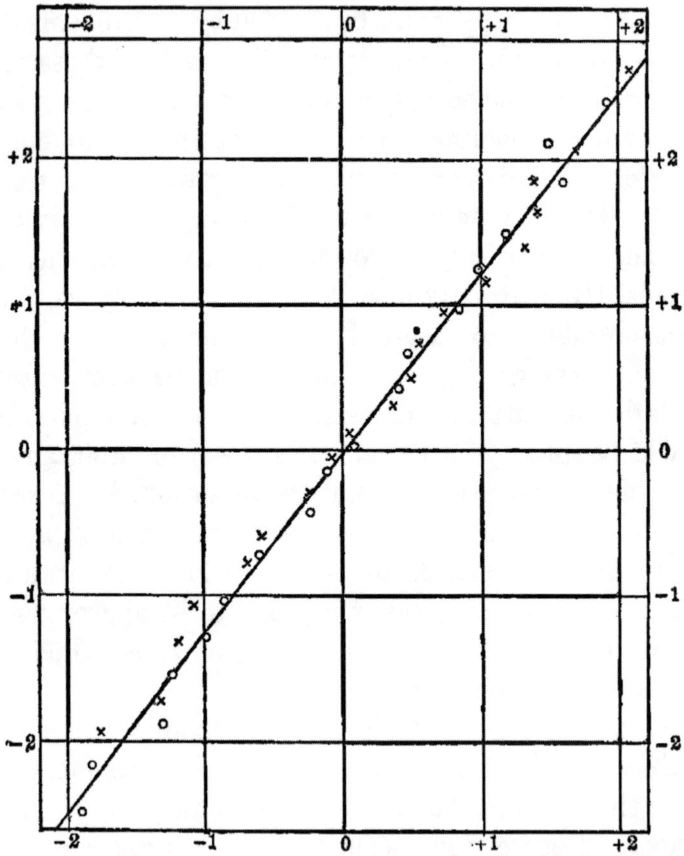

FIG. 16.—Correlation between post-spinous carapace length and total carapace length of shrimp.

measured along the abscissæ, they varying on an average from + 2.92 to — 2.17. The crosses in the diagram indicate the values obtained when the carapace length was fixed, and the circles those when the post-spinous portion was fixed. The line drawn through them indicates the ratio .81. Every point should theoretically lie on it, and it will be seen that they do, as a matter of fact, lie very closely around it.

This correlation constant of .81 was obtained by Professor Weldon for shrimps collected at a particular locality, viz., Plymouth. Similar determinations were also made for shrimps obtained from other localities, with the following results:

In Plymouth	$r = .81$	(1000 individuals examined)
In Southport	.85	(800 " ")
In Roscoff	.80	(500 " ")
In Sheerness	.85	(380 " ")
In Helder	.83	(300 " ")

The approach to identity between these values is very striking, the differences appearing to be within the probable error of each determination. There seems a reasonable ground for assuming, therefore, that the degree of correlation between the two particular organs measured is practically constant in all the races examined. The correlation between other organs was also estimated, but this was in each instance very much slighter, and in the case of the telson and sixth abdominal tergum, it was negative. Considering the degree of independence of these organs, as shown by the smallness of their correlation constants, the similarity between the values for the two local races is probably as close as could be expected. Hence, as both the organs measured and the samples of shrimps examined were chosen by chance, any result which holds for all these organs through all these races may be reasonably expected to prove generally true of all organs through the whole species.

	CARAPACE LENGTH AND TERGUM VI.	CARAPACE LENGTH AND TELSON.	TELSON AND TERGUM VI.
Plymouth,	.09	.18	—.11
Southport,	.06	.14	—.09

Professor Weldon points out that the above results lead us to hope that it may be possible to determine constants for any species of animal which would "give an altogether new kind of knowledge of the physiological connection between the various organs of animals, while a study of those relations which remain constant through large groups of species would give an idea, attainable at present in no other way, of the functional correlations between various organs which have led to the establishment of the great sub-divisions of the animal kingdom."

In a subsequent paper,* Professor Weldon determined no less than 23 different correlation constants, between various pairs of organs in 1000 adult female crabs (*Carcinus mœnas*), collected in Plymouth Sound, and in another 1000 collected in the Bay of Naples. He found that there was as a rule a remarkable degree of correspondence between the values of *r* derived from an investigation of the same pair of organs in the two races examined. There were in some cases considerable differences between the values, it is true, but Weldon considers that these were in no case sufficient to justify the assertion that the degree of correlation is really different in the two cases. It should be mentioned, however, that Professor Pearson † does not

* Proc. Roy. Soc., liv. p. 318.
† Phil. Trans. 1896, A. p. 267.

agree with this conclusion, but thinks that the differences observed are too large to justify such an assumption.

At Professor Weldon's suggestion, Mr. E. Warren * undertook similar measurements on 2300 specimens of another crab, *Portunus depurator*, also obtained from Plymouth. The accompanying table gives the results obtained by Warren, and some of those obtained by Weldon:

ORGANS.	C. MŒNAS, NAPLES RACE.	C. MŒNAS, PLYMOUTH RACE.	PORTUNUS DEPURATOR, PLYMOUTH.
Total breadth and frontal breadth,	.08	.10	.14
" " R. antero-lateral,	.66	.65	.67
" " R. dentary margin,	.50	.55	.56
Frontal breadth and R. antero-lat.,	.29	.24	.30
" " R. dentary margin,	−.23	−.18	−.03
" " L. dentary margin,	−.26	−.20	−.01
R. antero-lateral and L. antero-lat.,	.76	.78	.86
" " R. dentary margin,	.71	.78	.80
" " L. dentary margin,	.60	.70	.74

On glancing through this table, it will be seen that, with two exceptions, the values for the two races of *C. mænas* differ from one another nearly as much as they do from the constants of Portunus, an animal belonging to a different genus. It is probable, however, that the larger differences in this latter animal do indicate real differences in the correlation constants, associated, perhaps, with changes in habit or environment. For example, it is conceivable that a crab which swims might find it advantageous to be more symmetrical than one which only crawls between the tide marks. Portunus does swim to a certain extent, and we see that the correlation of the two sides of its body is greater than in the essentially shore-living Carcinus.

* Proc. Roy. Soc., lx. p. 221.

Another most laborious research undertaken at Professor Weldon's suggestion is that of H. Thompson,* on the correlation of certain external parts of the prawn, *Palæmon serratus.* Twenty-two measurements were made on 1000 adult females, and from these the value of Galton's function was calculated for 56 pairs of organs. As might be expected, the degree of correlation was highest between the paired organs; *e. g.*, .94 between the right and left squames. Also there was a strong correlation between the terga of adjacent abdominal segments, their values ranging between .58 and .71.

Of other recent work on correlation, that by Miss Lee and Professor Pearson † may be briefly alluded to. This consists in a comparison of measurements on certain long bones of about 40 male and 25 female skeletons of the Aino race (a primitive tribe dwelling in Japan), with corresponding measurements of 50 male and 50 female skeletons of the modern French race. It was found that the transition from the uncivilised to the civilised condition is accompanied by well-marked changes in the sexual relationships; primitive man and woman being more nearly equal in size, variability, and correlation than highly civilised man and woman. Civilised man has gained in size on woman, but this has been accompanied by a relative loss in variability and the correlation of parts. The general result of increased civilisation is to increase the absolute size and amount of variation. In females, also, the degree of correlation is increased, but in males this remains sta-

* Proc. Roy. Soc., lv. p. 234.
† Proc. Roy. Soc., lxi. p. 343.

tionary. It is therefore impossible to say that civilised woman is nearer to the primitive type than civilised man, for while civilised man differs more from the primitive type than civilised woman, so far as absolute size is concerned, he has made only about half her progress in variation, and hardly any progress in correlation. The absolute amount of correlation is very high, as the following figures show:

ORGANS.	MALES. AINO.	MALES. FRENCH.	FEMALES. AINO.	FEMALES. FRENCH.
Femur and tibia,	.83	.81	.85	.89
" humerus,	.86	.84	.87	.87
" radius,	.79	.74	.70	.78
Clavicle and humerus,	.44		.63	
Humerus and radius,	.78	.85	.74	.85
Tibia and fibula,	.89	.96	.97	.98
Humerus and ulna,	.77	.77	.75	.86
Radius and ulna,	.98	.88	.98	.92

Here we see that the Galtonian constant, r, was in most instances above .8. Between the tibia and fibula it averaged .95, and between radius and ulna .94, so that in these cases the correlation was almost absolute.

In the Naquada race investigated by Warren * the correlation between the lengths of the long bones in males (as measured in about 60 skeletons) was distinctly higher than for Aino males, but in females (measured in about 90 skeletons) it was either the same or was lower than in Aino females.

Probably correlation is to some extent affected by sex in most animals. Thus Duncker † determined the correlation coefficients of 40 pairs of measurements in male

* *Vide* Phil. Trans. 1898, B. p. 178.

† Wissenschaftliche Meeresuntersuchungen aus der biologischen Anstalt auf Helgoland, Bd. iii. p. 351.

and female flounders (*Pleuronectes flesus*), and the values obtained showed that the correlation was affected by sex in 17 out of the 40 instances. The coefficients were greater in the male than in the female fish in 11 instances, and in the female than in the male in 6 instances. Several of the pairs of bilateral homologous measurements (such as the numbers of rays in the right and left pectoral and ventral fins) showed distinctly lower correlation constants than were shown by the corresponding pairs of measurements in the symmetrical fish *Acerina cernua* and *Cottus gobis.* This was doubtless due to their possessing slight differences of function in the asymmetrical fish.

It should be mentioned that Pearson, Warren, and Duncker employed a somewhat modified and improved formula for determining these correlation constants,* as compared with that originally suggested by Galton. G. O. Yule,† and also Pearson and Filon,‡ have recently shown that the correlation can be determined in the case of skew variation, as well as of normal variation.

All these results may be taken to show that every part and organ of the body is correlated with every other part in a greater or less degree, though such correlation may sometimes be of the negative order. The immense importance in evolutionary processes of such correlation, whereby when one organ becomes modified by the action of an agency such as Natural Selection, others are modified also, is sufficiently obvious to need no discussion.

* Phil. Trans., 1896, A. p. 264.
† Proc. Roy. Soc., lx. p. 477, 1897.
‡ Phil. Trans., 1898, A. p. 229.

Besides these instances in which the degree of correlation can be expressed in numerical equivalents, there remain other cases in which such expression is difficult or impossible. In these it must accordingly be defined in general terms. Darwin has collected a large number of such cases in his "Animals and Plants," * but it is not necessary to reproduce more than a few of the most striking of them here. For instance, Tegetmier has stated that young pigeons of all breeds which when mature have white, yellow, silver, blue, or dun-coloured plumage, are born almost naked; whereas pigeons of other colours are clothed with plenty of down. Darwin himself has noticed that in feather-footed pigeons, not only does the exterior surface support a row of long feathers, like wing feathers, but the very same digits which in the wing are completely united by skin become partially united by skin in the feet. Again, Polish fowls have a large tuft of feathers on their heads, and their skulls are perforated by numerous holes. That this deficiency of bone is in some way connected with the tuft of feathers is clear from the fact of tufted ducks and geese likewise having perforated skulls. Constitutional peculiarities are sometimes correlated with colour in a most curious and interesting manner. For instance, Beddoe has shown that a relation exists between liability to consumption and the colour of the hair, eyes, and skin. As regards animals, white terriers suffer most from distemper, white chickens from a parasitic worm in their tracheæ, white pigs from scorching in the sun, and white cattle from flies. Again, all the hogs in Virginia, excepting those

* Chap. xxv.

of a black colour, suffer severely from eating the root of *Lachnanthes tinctoria*. Similarly, buckwheat when in flower is highly injurious to white or white-spotted pigs, if they are exposed to the heat of the sun, but quite innocuous to black pigs.

These few instances suffice to show how widespread and apparently capricious may be the range of correlation. Until careful observations have been made, accompanied when possible by measurements, one can never on *a priori* grounds assume that there is no correlation between particular organs or parts of an organism, and that an agency acting on one part may not at the same time be thereby indirectly modifying another.

In their effects on the modification of species, probably by far the most important cases of correlation are those in which the reproductive system is concerned. Until recently comparatively little attention was paid to such phenomena, and probably, even now, they are far from being estimated at their true value by many biologists. Perhaps the reason of this lies in the fact that the physiological condition of an organism, or its relative degree of sexual compatibility with other organisms of its own and of different species, is so exceedingly difficult, if not impossible, to estimate. Thus the degree of fertility cannot be tested in more than one or two instances with each individual organism—at least in the higher animals—and so the desired information can only be acquired by carrying out most lengthy and laborious series of observations. There are sufficient data at our disposal, however, to indicate that the reproductive system is no less subject to variation than any other part of the organism, and indeed is probably a good deal more so. Supposing that the quality of fertility is correlated with some particular character or characters more than it is with other characters, then it follows that more individuals bearing the character in question will be born and propagate their kind, and so, in course of time, the whole race will be modified in this direction. This principle has been termed by its discoverer, Professor Pearson, "Reproductive" or "Genetic Selection." Its existence as a real factor in evolution depends on the validity of the assumption that the characteristic of fertility is inherited. That this is so, Professor Pearson, in conjunction with Miss Alice Lee and Mr. L. Bramley-Moore,* has recently proved by a most laborious research on inheritance in man and in the thoroughbred race-horse. Their results show that fertility is undoubtedly inherited from mother to daughter, and also from father to son. It was also found that a woman's fertility is as highly correlated with that of her paternal as with that of her maternal grandmother. In other words, the latent character fertility in the woman is transmitted through the male line, and with an intensity which approximates to that required by the law of ancestral heredity. Again, it was deduced that fecundity in the brood-mare is inherited from dam to mare, and also from grand-dam to mare through the dam. Also the latent quality of fecundity in the brood-mare is inherited through the sire, and by the stallion from his sire. In these latter two cases, the degree of inheritance approaches fairly closely to that required by Galton's law of ancestral heredity, but, in the two for-

* Proc. Roy. Soc., lxiv. p. 163, 1899.

mer, it is much less. As, therefore, fertility is proved to be inherited in man, and fecundity in the horse, it is probable that both these characters are inherited in all classes of life.

The importance of this theory of Genetic Selection will, perhaps, be better realised by quoting a concrete case concerning man, this being the only one in which statistics are at present available. Working on data concerning 4000 families, principally of the Anglo-Saxon race, and 1842 families of the Danish race, Professor Pearson * determined that there is a sensible correlation (about .18) between fertility and height in mothers of daughters. Supposing genetic selection to have been unchecked by natural selection, say for forty generations, the mean height of women would have been raised about $3\frac{1}{4}$ inches. A factor which would alter stature by about three inches in 1000 years is clearly capable of producing very considerable results in the long periods during which evolution may be supposed to have been at work. The importance of the influence of genetic selection in the case of man is also shown by the fact that, as proved by these statistics, less than a quarter of one generation, by reason of their fertility, produce more than half of the next generation. Correlation between fertility and any mental or physical characteristic must therefore work a progressive change. For example, arguing from the class fertility statistics which have been determined among the population of Copenhagen, it is gathered that the artisan class produce a larger proportion of children than the professional classes, as their gross fertility is greater, and

* Proc. Roy. Soc., lix. p. 301, 1896.

their marriage rate is so much higher. This increased fertility is somewhat counteracted by their greater death rate, but it would nevertheless appear that the population will ultimately reproduce itself from the artisan classes.

Definite evidence of evolution under natural conditions as the result of genetic selection has not been obtained, but this may be simply because it has never occurred to anybody to look for it. Professor Pearson * has, however, shown the existence of a highly interesting and important relationship between fertility and morphological characters in certain plants. He counted the number of stigmatic bands on the 4443 seed-capsules obtained from 176 Shirley poppies growing in a single garden, and found the following frequencies:

Bands,	5	6	7	8	9	10	11	12	13	14	15	16	17	18	19
Frequency,	1	11	32	56	148	363	628	925	954	709	397	155	51	12	1

Here we see that the 12 and 13 band forms were the most common, the 11 and 14 ones less so, the 10 and 15 still less, and so on. To his surprise, Professor Pearson found that whilst the commonest or type capsules contained a very large number of seeds, the 11 and 14 band forms contained distinctly less, and the 10 and 15 ones very few seeds indeed, whilst the capsules with very few or very many bands contained practically no seeds. A repetition of these observations on the wild poppy gave a very similar result, and this was likewise the case with the seed capsules of a number of plants of *Nigella Hispanica*. The distribution of the segmentation on 3212 capsules was as follows:

* "Grammar of Science," Ed. ii. p. 443.

No. of Segments,	2	3	4	5	6	7	8	9	10	11	12	13	14	15	16	17	18	19	20
Frequency,	10	7	20	303	412	534	1552	223	59	35	43	6			6				2

In this plant Pearson found the 8 segment capsules to be highly fertile, whilst in the 10, 11, and 12 segment capsules he could find hardly any seed at all. Six and 7 segment capsules were only moderately fertile, and those with 5 segments or less were practically sterile. These experiments Professor Pearson holds to illustrate a very important law, namely, "Fertility is not uniformly distributed among all individuals, but for stable races there is a strong tendency for the character of maximum fertility to become one with the character which is the type." It follows, therefore, if this principle is generally true, that stable races are very largely the product of the typical or most frequently occurring members, and not of all the individual members in proportion to their numbers. There would seem to be a constant tendency to keep the type uniform and limit the variability in either direction as much as possible. Doubtless under changed conditions of environment, the relative fertility might also become changed, and in consequence a gradual evolution result.

A similar relationship between fertility and type form has been noticed by Davenport * in one of the Hydromedusæ, *Pseudoclytia pentata.* This organism differs from all other Hydromedusæ in that it normally has five radial canals, instead of four. A. G. Mayer † has examined the variation of the species, and obtained the frequencies given in the subjoined table. From these values we gather that the four and six canal forms

* Biometrika, i. p. 255, 1902.

† Science Bulletin of the Brooklyn Museum, vol. i.

are comparatively common, whilst other abnormalities are rare. Typically one gonad (or reproductive organ) occurs on each radial canal, but on an average about one in seven of them fails to develop. In atypical forms, however, as can be gathered from the table, the proportion failing to develop is found to be larger and larger, the further the departure of the individual from the type. In such forms as depart from the normal radial symmetry, even if they still possess five rays, the partial sterility is very much increased.

Number of Canals.	Frequency of Occurrence.	Per Cent. of Gonads Failing to Develop. All Individuals.	Irregular Individuals.
2	1		
3	8	25.0	44.5
4	56	21.8	31.0
5	860	15.6	37.0
6	64	18.6	31.0
7	6	21.5	43.0
8	1		

A very interesting case of variation described by Bateson * may perhaps be ascribed to the action of genetic selection, though there is no direct evidence to warrant the assumption. It concerns the Peppered moth, *Amphidasys betularia.* A striking black variety of this insect, *A. doubledayaria,* was first met with as a rarity in 1840-50. Since then its numbers have gradually increased, till in 1870 about equal numbers of the pure type and of its variety occurred at Monmouth, whilst a few years later the typical form had entirely disappeared. At Chester none but black forms have been met with for many years. In the south of England, however, the typical form is still alone present. Bate-

* Science Progress, vol. vi. p. 557, 1897.

son suggests that this gradual replacement of a type by its variety is probably due to success in the struggle for existence of this particular dark strain, but it may equally well be accounted for by supposing that it is the result of a greater fertility. Intermediate strains are not unknown, for in Belgium it seems clear that one has succeeded in establishing itself, and in England it is probable that they were once more common than they are now. Intermediate forms are said to be plentiful also in the Rhenish Provinces and Westphalia, and the same is true of the black forms. Bateson says there is no doubt that the black variety existed at an early stage in the transformation, side by side with the light one. The course of events has not been that the insects of each successive district have become more and more tinged with black, till they culminated in *A. doubledayaria*, but rather that this variety, or less often one of the intermediate forms, spread into or at least appeared in the area, and either coexists with the type or has replaced it.

The few breeding experiments thus far made on the moth show that there is an imperfect blending of type and variety. Steinert raised from a black wild female a brood of 75 typical and 90 varietal forms, but there were no really intermediate ones, though two of the examples classed as *betularia* were darker than the normal. That this female had paired with a typical male —the progeny thus resembling either the one parent or the other, but not both—is shown by the following experiment, also recorded by Bateson.* W. H. B. Fletcher tied out a black female, which had been reared

* Science Progress, vol. vii. p. 53, 1898.

in captivity. This was at Worthing, where the typical male form is the only one known, so that the brood obtained were most certainly produced from a cross of the two varieties. The offspring were sharply divided into 10 male and 8 female *betularia*, and 6 male and 5 female *doubledayaria*.

Conversely to the modification of a community in one direction by reason of the increased productiveness of certain of its members, we may experience modification in the opposite direction by reason of a decreased productiveness. For instance, Dr. Beddoe * states that there is a good deal of evidence as to the greater liability of blonds than of brunets to certain classes of disease. At least this is so in America, as has been shown by Baxter.† Thus it would appear that the blonds have less chance than the brunets of contributing their due proportion to the next generation, and so they must be relatively diminishing in numbers. That this is the case is supported by the fact that of Americans accepted for service in the army a greater proportion were brunets than of the English, Irish, and Germans accepted. Thus:

Among	the	Americans	were	66	light	and	34	dark complexioned
"	"	English	"	70	"	"	30	" "
"	"	Irish	"	70	"	"	30	" "
"	"	Germans	"	69	"	"	29	" "

The fact that most species are to some extent mutually sterile, whilst their hybrid offspring are almost invariably so, proves that the physiological condition of

* Science Progress, vol. v. p. 384, 1896.

† "Medical Statistics of the Provost-Marshal-General's Bureau," Washington, 1875.

the reproductive system is closely correlated with the physiological condition of the organism taken as a whole. Recent physiological research has taught us that most organs in the body, in addition to their more obvious functions, have an internal secretion which passes into the circulation of the body, and there exerts some important, but unexplained, influence on the general metabolism of the tissues. Deprivation of such internal secretion, by extirpation of one of these organs, may speedily upset the normal working of many or most of the other tissues of the body, and ultimately result in death. Every organ and tissue of the body probably reacts on every other organ, and modifies its physiological condition, and thereby may ultimately produce structural changes. The reproductive system is apparently more sensitive than most other organs, and hence is very readily affected by any changes in the condition of the organism as a whole. For instance, it is well known that most animals refuse to breed in confinement, though they can be kept for many years in a condition of perfect health. The changed conditions of life must therefore have acted on the organism as a whole, so as to modify certain of its internal secretions, and these, reacting on the reproductive system, have brought about the observed sterility. If a considerable change in conditions of life produces complete sterility, it seems highly probable that slight changes in such conditions may produce a partial sterility, or a differential fertility. Thus organisms in a state of nature, if exposed to a change of climate, the result of migration or stress of weather, or to a change in their food, may have the physiological condition of their reproductive system slightly altered, whereby the principles of Genetic Selection and—in the case of plants—of Physiological Selection, may become effective, and modify or split up the species.

Just as the condition of the organism as a whole may modify that of the reproductive system, so may the condition of the reproductive system modify that of the organism. The difference between the spirit and appearance of castrated animals and that of normal animals is sufficiently well known to need no remark. Such differences must be due in large part to the lack of internal secretion from the organs of reproduction. Striking as is the influence of the reproductive organs on the physiological condition of an animal, that upon the morphological structure is even more noteworthy. That castrated male animals fail to develop their secondary sexual characteristics, is notorious. Thus, if the operation be performed upon a young cock, he never crows again; the comb, wattles, and spurs do not grow to their full size, and the hackles assume an intermediate appearance between true hackles and the feathers of the hen. Conversely, it is well known that a large number of female birds, such as fowls, various pheasants, partridges, pea-hens, and ducks, when old or diseased, or when operated upon, assume many or all of the secondary male characters of their species. Waterton gives a curious case of a hen which had ceased laying, and had assumed the plumage, voice, spurs, and warlike disposition of the cock. Again, the females of two kinds of deer, when old, have been known to acquire horns.*

"Animals and Plants," vol. ii. p. 26.

Analogous or Parallel Variation. This term has been used by Darwin to indicate that similar characters occasionally make their appearance in several varieties or races descended from the same species, and more rarely in the offspring of widely distinct species.* It is unnecessary to make more than very brief mention of this subject, because, as Darwin points out, the majority of observed cases—such as the occasional appearance of black wing bars in the various breeds of pigeon, and of stripes on the legs of the ass and of various races of the horse—are evidently due to reversion. The others are probably mere coincidences, and of no scientific value. Among these latter, Darwin mentions the fact that many trees belonging to quite different orders have produced pendulous and fastigate varieties. A multitude of plants have yielded varieties with deeply cut leaves. Several varieties of melon resemble other species in important characters. Thus one variety has fruit so like, both externally and internally, the fruit of the cucumber, as hardly to be distinguished from it. In animals, again, we find feather-footed races of the fowl, pigeon, and canary bird. Horses of the most different races present the same range of colour. Many sub-varieties of the pigeon have reversed and somewhat lengthened feathers on the back parts of the head.

In connection with this subject of parallel variation, Walsh's "Law of Equable Variability" † may be mentioned. This states that "if any given character is very variable in one species of a group, it will tend to be variable in allied species; and if any given character

* "Animals and Plants," vol. ii. p. 340.

† Proc. Entomolog. Soc. Philadelphia, p. 213, 1863.

is perfectly constant in one species of a group, it will tend to be constant in allied species." The general truth of this law seems highly probable on the face of it, because most allied species have presumably split off from their common ancestor at no very remote period, and so would still resemble each other more or less closely in respect of variability, just as they do in respect of morphological structure. The results, already quoted, of Weldon and Warren for the correlation of various organs in Carcinus and Portunus, afford direct experimental evidence in support of this law.

In conclusion, it may not be out of place to make one or two brief remarks as to the general bearing of the evidence which has been adduced in these three chapters concerning the "facts of variation." I believe that they include most of the more important and more recent contributions to our knowledge of the subject, especially those in which the results have been expressed in exact numerical terms. It may very likely be objected that insufficient importance has been attached to the kind of information which Darwin collected so thoroughly and in such profusion in his work on "Variation in Animals and Plants under Domestication."

The reason of this is twofold. In the first place it seemed unnecessary to refer at great length and with much frequency to data with which most serious students of Biology must be already conversant: whilst in the second place, comparatively little information of this kind has, as far as I am aware, been recorded in scientific journals since Darwin's time. It has, indeed, been recognised that the facts of variation attain a much higher and more permanent value in proportion as they are expressed in exact numerical terms.

To say that some organism or part of an organism is more variable than another is very unsatisfactory, compared with the statement that the variabilities of certain characters in the one are of such and such values, and in the other, of certain other values. From such data as these we can compare the variabilities of all the variants exactly, both with each other and with any other variants, and determine what relation, if any, they bear to their systematic importance. We can tell if the variations obey the normal law of error, or if they are asymmetrical in their distribution. In this latter case, we may be able to discover the existence of a tendency to divergence or splitting up of a species, in its very earliest stages. Repetition of our observations at some future period would thus become a subject of especial interest, as we might in such a case hope to detect some change both in the mean values and in the distribution of the variations, indicating that the evolutionary process was still progressing, and the direction in which the progress was being made. Further, as we shall see in a subsequent chapter, by determining the average characters of groups of individuals which have been subjected for some period to the struggle for existence, and comparing them with the characters of other individuals which have not been exposed to such a struggle, or by comparing the characters of individuals, which owing to the severity of the struggle for existence had been actually eliminated, with the characters of the survivors, we are able to obtain numerical proof of the action of Natural Selection. Still again, by comparing the characters of individuals with those of their parents and of their offspring, we are enabled to work out with exactness the origin and transmission of such characters, and so elucidate the Laws of Heredity.

PART II.

THE CAUSES OF VARIATION.

CHAPTER IV.

BLASTOGENIC VARIATIONS.

The ultimate cause of blastogenic variation—Effect of staleness and of comparative maturity of sex-cells on the characters of organisms—Amphimixis—Identical twins—Transplantation of ova in the rabbit—Law of Ancestral Heredity in man and in the Basset hound—Regression towards mediocrity—Exclusive inheritance.—Homotyposis.

ARGUING from his theory of the continuity of the germ-plasm, first suggested in 1883,* Weismann came to the conclusion that acquired characters were not transmissible. Such acquired characters are due to the direct influence of the environment upon the body tissues of an organism, or are variations of *somatogenic* origin. Opposed to these are variations due directly to certain peculiarities of the germ-plasm, or variations of *blastogenic* origin, which are, on the contrary, hereditary or transmissible. Thus, according to Weismann, all variations are, in respect of their origin, sharply divisible into these two groups, whilst in respect of transmissibility they are equally distinct. In the present

* *Vide* "Essays on Heredity," Oxford, 1889, p. 71.

chapter these blastogenic, genetic, or germinal variations will be discussed, whilst somatogenic variations will be treated of later.

The cause of hereditary variation Weismann ascribes to the direct effect of external influences on the so-called biophors and determinants of the germ-plasm.* These biophors Weismann defines as "bearers of vitality," or the smallest units of protoplasm which exhibit the primary forces of assimilation and metabolism, growth, and multiplication by fission. All protoplasm, both the nucleus and body of cells, is made up of these units. They are the bearers of the qualities or characters of the cells. Determinants, on the other hand, are groups of biophors, and are the particles of germ-plasm corresponding to and determining the cells or groups of cells which are independently variable from the germ onwards.

The reaction of the germ-plasm to external influences is primarily one of nutrition. The biophors and determinants are supposed to be subject to continual changes of composition during their almost uninterrupted growth, and these very minute fluctuations are the primary cause of the greater deviations of the determinants, which are finally observed in the form of individual variations. The growing determinants must originally differ to some slight extent in the composition of their biophors, as otherwise inequalities of nutrition could never effect any transformation, but could only alter their rate of growth. Slight as are these deviations in the determinants effected by inequalities of nutrition, they are nevertheless of great significance,

* "The Germ-Plasm," London, 1893, p. 415.

as, by a process of accumulation, they form the material from which the visible individual variations are produced by means of "amphimixis."

Amphimixis is that form of reproduction which is found in all the higher organisms, and which consists in the mingling of two individuals or their germs; *i. e.*, the so-called sexual reproduction. The term is also applied to a similar phenomenon occurring amongst unicellular organisms, *i. e.*, to conjugation. In this case reproduction is not a necessary or even usual concomitant, but takes place independently. To amphimixis Weismann attributes the constant occurrence of individual variability, although he recognises that it is not the primary cause of this variability; but rather the process furnishes an inexhaustible supply of fresh *combinations* of individual variations. Thus the germ-plasm of a new individual produced by amphimixis never receives more than half the ids of each parent, and these are differently selected and arranged in each case. By an *id*, it may be remarked, Weismann means a group of determinants which contains all the determining elements of the species, though in a manner peculiar to the individual.

Blastogenic variations are thus, according to Weismann, primarily dependent on two chief factors: (1) Inequalities of nutrition acting on the individual constituents of the germ-plasm; (2) Amphimixis.

It behoves us to examine these two factors more closely, and see how far they are supported by experimental evidence. It is, on the face of it, impossible to put Weismann's hypothesis of the reaction of determinants to inequalities of nutrition to a practical test, but we can at least enquire into what is known about the influence of nutrition on the germ-plasm as a whole. In fact, we can see how far the parental plasms are individually capable of being affected by changes of nutrition, so as, on subsequent mingling in sexual union, to give rise to appreciable changes in the resulting offspring. Important as this subject is, the direct experimental evidence available upon it is distinctly meagre. It is, for instance, probable that the children of a father whose tissues, and therefore his sex-cells, are saturated with alcohol or the products of some disease, are smaller and less well formed than those of normal parents, but there are no satisfactory data to support it. Similar evidence with reference to the female sex-cells is obviously not available, as any effects produced on offspring would probably in chief part arise during embryonic development, or after, and not before, fertilisation.

The evidence obtained as to the influence of nutrition on the evolution of sex is only indirectly related to the problem under discussion. Most of it goes to show that increased feeding of young organisms tends towards the production of a larger proportion of females,* and hence, as male and female sex-cells cannot be considered entirely equivalent, it follows that an effect is produced on the germ-plasm. Yet there is no evidence to show that the offspring of females which arose in spite of bad feeding differ in any way from those of females produced in consequence of good feeding. Nevertheless it seems probable that, as nutrition has some influence in determining the sexual

* *Vide* Geddes and Thomson's "Evolution of Sex," p. 41, 1889.

character of the germ-plasm in a developing animal after fertilisation, it may also have some influence if it be brought to bear on the parental germ-plasms before fertilisation. It seems likely, in fact, that a highly nourished ovum, as compared with one less favourably conditioned, will tend rather to a female than a male development.

Evidence bearing more directly on the question at issue has recently been obtained by the author, in a research on the effect of staleness of the sex-cells on development.* The method of experiment was to keep the ova or spermatozoa, or both, of the sea-urchin *Strongylocentrotus lividus* for varying numbers of hours in sea water before permitting fertilisation, and after eight days' development to measure the length of the larvæ and see if they differed in size from normal larvæ. As other observations on larvæ obtained from these artificial fertilisations will be referred to later, the experimental procedure adopted may be briefly indicated. It consisted in shaking pieces of the ovaries and testes of several ripe specimens of the Echinoid in small jars of water, and mixing portions of their contents either immediately, or after a given number of hours. The mixed solutions were allowed to stand for an hour, and were then poured into large jars holding from 2 to 4 litres of sea-water. These were placed in a tank of running water, whereby the temperature was kept nearly constant, it varying less than a degree during twenty-four hours and not more than two degrees during the whole course of the experiment. The fertilised ova were allowed to develop for eight days, as by

* Proc. Roy. Soc., vol. lxv. p. 350, 1899.

that time the arms of the larvæ or plutei have attained their maximum length, whilst the body has practically ceased growing. The larvæ were then killed by adding corrosive sublimate. They were collected and preserved in 80 per cent. alcohol, and were subsequently mounted in glycerine and measured under the micro-

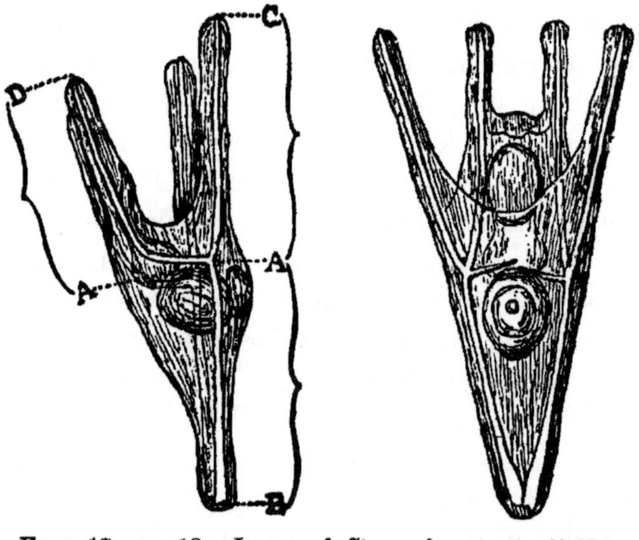

FIGS. 17 AND 18.—Larvæ of *Strongylocentrotus lividus.*

scope by means of a micrometer eyepiece. The body-length, AB, was always measured in 50 different larvæ, and a mean taken. In many cases also the anal arm-length, AC, and sometimes also the oral arm-length, AD were measured as well, and these measurements calculated as percentages on the body length.

In order to determine the effect of staleness of the sex-cells on the size of the larvæ, five series of measurements had to be made, viz., (1) of the normal larvæ obtained from the fresh ova fertilised with the fresh sperm, (2) those from stale ova and stale sperm, (3) from stale ova and fresh sperm obtained from another freshly opened Echinoid, (4) from fresh ova and stale sperm, (5) and lastly, from the ova and sperm of the freshly opened Echinoids. It was, of course, impossible to get an exact basis of comparison for the larvæ obtained from one stale and one fresh sexual product. The best possible was to take a mean between the size of the original normal larvæ, and that of the larvæ obtained from the fresh sexual products used for fertilising the stale products. The larvæ obtained with both sexual products stale are, of course, accurately comparable with the original normal larvæ. In the accompanying table an example is given of the mean percentage differences in the size of the larvæ obtained with fresh and stale products, from the original normal larvæ in the one case, and from the mean between the original and fresh normal larvæ in the other two

CONDITION OF SEXUAL CELLS.	FERTILIZATION MADE AFTER			
	9 HRS.	24 HRS.	33 HRS.	45 HRS.
Stale ♀, stale ♂	−0.2	+1.9	+ 1.1	− 1.9
Fresh ♀, stale ♂	+7.1	+3.7	+10.9	+ 1.5
Stale ♀, fresh ♂	−2.8	−3.0	+ 2.0	−15.9

cases. In this experiment, which was the most complete made, the fertilisations were performed after keeping the sexual products for respectively 9, 24, 33, and 45 hours. It will be seen that the larvæ obtained when both sexual cells were stale were of practically the same size as when they were fresh, even after they had been kept 33 or 45 hours. The larvæ from fresh ova and stale sperm, on the other hand, were in each case *distinctly larger* than the normal, they differing on an average by + 5.8 per cent., whilst those from stale ova and fresh sperm were *distinctly smaller*, they differing by — 4.9 per cent. In one of these latter observations there was, for some unknown reason, a slight increase in size, but there can be no doubt that on the whole the tendency was towards diminution. On taking means of all the values obtained in this and in other similar experiments, it was found that as an average of eight observations, the stale ♀ stale ♂ larvæ were diminished by .7 per cent. in size; as an average of eleven observations, the fresh ♀ stale ♂ larvæ were increased by 4.0 per cent., and as an average of ten observations, the stale ♀ fresh ♂ larvæ were diminished by 6.9 per cent. There can be no doubt, therefore, that variations in the degree of freshness of the sexual cells, that is to say, in the comparative state of nutrition of the germ-plasm as a whole, do have a very appreciable effect upon the size of the subsequently developing larvæ. It is to be particularly noticed that the effect produced differs entirely according to the sex-cells acted upon, and hence affords distinct evidence of the possibility that different portions of the same sex-cell may also react differently to one and the same change of nutrition.

Perhaps a more convincing proof of the influence of the nutritional condition of the sex-cells on the offspring they produce is afforded by certain other observations on these larvæ. On two separate occasions * series of

* Phil. Trans. 1895, B. p. 585, and 1898, B. p. 483.

artificial fertilisations were carried out at short intervals over periods of several months, and the larvæ allowed to develop under conditions which were probably nearly constant, except as regards temperature. For this varying factor a correction could be easily applied. In spite of the constancy of environmental conditions, however, the size of the larvæ showed very marked

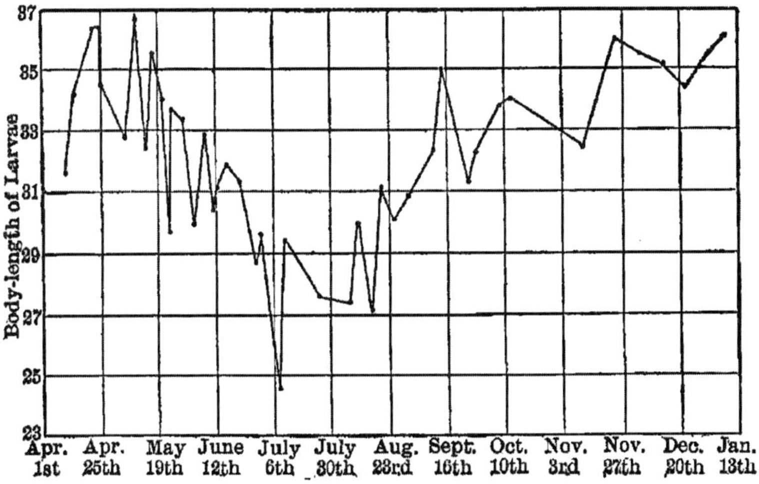

FIG. 19.—Seasonal variation in size of larvæ.

variations. The range of these variations may be gathered from the accompanying diagram. Here the ordinates represent the mean body lengths of the larvæ in micrometer eyepiece units (of which 152.3 are equivalent to 1 mm.), and the abscissæ the time of year at which the fertilisations were made. It will be seen that in April and May the larvæ were on an average about 34 units in length, but that then they began steadily to dwindle down in size, so that in June they were about 31 units, and in July and August only about

28 units. From this minimum they then rose rapidly, so that in September they were about 32 units, in October and November 34 units, and in December and January 35 units in length. The extreme variations were from 36.80 to 24.49, the larvæ of this latter length, obtained on July 9, being no less than 33.4 per cent. smaller than the former.

This extraordinary seasonal variation in the size of the larvæ is probably very closely, if not entirely, dependent on changes in the maturity and nutritional condition of the sexual products. Thus, of the specimens of *Strongylocentrotus* obtained in the winter months, almost every individual contained ripe sexual products in large quantities, whilst of those obtained in the summer months, not more than about one in four yielded any ova or sperm at all on shaking the ovaries and testes in water, and occasionally twenty or more individuals were opened before any ripe sperm was obtained. Again, the best of the specimens obtained in the summer months did not contain nearly so much of the ripe sexual products as they did in the winter.

That this effect of season on the condition of the sex-cells is more far-reaching than is implied in a mere diminution of size in the resulting offspring, is proved by some observations on the crossing of this species of sea-urchin with another species, viz., *Sphærechinus granularis*. Hybrids between *Sphærechinus* ova and *Strongylocentrotus* sperm can probably be obtained, though it may be only after several attempts, at all times of the year. It was found, however, that their structure was by no means constant.* The majority of

* Phil. Trans. 1898, B. p. 470.

the hybrids obtained in May, June, and July were of the almost pure *Sphærechinus* type, of which an example is given in Fig. 20; but about a third of them or less were of the intermediate or *Strongylocentrotus* type, of which an example is given in Fig. 21. In November, on the other hand, only about a sixth of the hybrid larvæ were of the *Sphærechinus* or maternal type, and

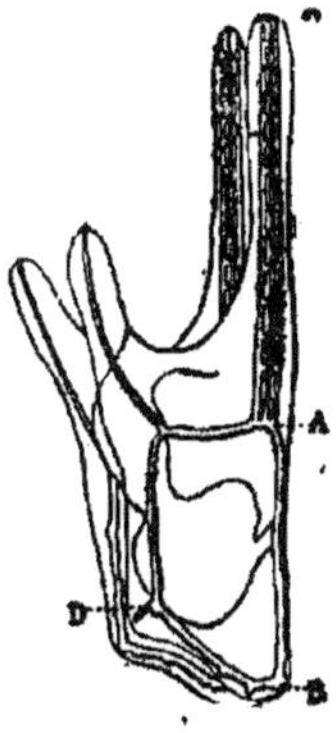

FIG. 20.—*Sphærechinus* larva.

FIG. 21.—*Sphærechinus* ♀ *Strongylocentrotus* ♂ larva.

five-sixths of the paternal type. Finally, in December and January, all the larvæ were of the paternal type. These so-called paternal larvæ in almost all cases showed obvious traces of their hybrid origin, but they were evidently much more inclined to the *Strongylocentrotus* than to the *Sphærechinus* type.

Combining this series of observations with that just recorded, we therefore find that in the summer months, when the *Strongylocentrotus* sperm is in a condition of minimum maturity, the *Sphærechinus* ♀— *Strongylocentrotus* ♂ hybrids are chiefly of the *Sphærechinus* type. As, however, the maturity of the sperm increases, it is able to transform first a portion and then the whole of the hybrid larvæ from the *Sphærechinus* to the *Strongylocentrotus* type. A repetition of these crossing experiments in a subsequent year * confirmed the conclusion that the summer hybrids were more inclined to the *Sphærechinus* type than the winter ones, though on this occasion they were only very rarely found to approach to the pure *Sphærechinus* type.

The reciprocal cross of *Strongylocentrotus* ova with *Sphærechinus* sperm illustrates still another way in which the sex-cells may be affected by changes in maturity and nutrition. Thus during April, May, and June a fair number of the ova were cross-fertilised, though no plutei were obtained: but in July and August some 47 per cent. of the ova were fertilised, and 29 per cent. of them survived to the eight days pluteus stage. In November and December, on the other hand, with one exception, not only were no plutei obtained, but, as a rule, not a single ovum was cross-fertilised. In other words, the *Strongylocentrotus* ♀ *Sphærechinus* ♂ hybrid is only formed at a time when the *Strongylocentrotus* ova have reached their minimum maturity.

The observations made upon these sea-urchin larvæ thus afford conclusive evidence that changes in the conditions of nutrition of the sex-cells produced by keeping them in sea water may affect the size of the larvæ both in a positive and negative direction, whilst changes of nutrition dependent on season may produce a much more considerable effect on size, and may in some instances so alter the nature of the germ-plasm as to give rise to most marked variations of structure in the resulting hybrid offspring, and in other instances largely

* Arch. f. Entwickelungsmechanik, Bd. ix. p. 464, 1900.

to abolish the normal resistance to cross-fertilisation. If differences of nutrition in the parental germ-plasms as a whole can produce such profound effects on the offspring to which they give rise, then, supposing it is possible that various individual portions of the germ-plasm are capable of being more or less independently affected by inequalities of nutrition, there seems no reason to doubt that they may show a similar reaction, and so give rise to variations in the individual parts of an organism which they represent or "determine."

Experiments upon higher organisms are, of course, very much more difficult to carry out than those upon sea-urchin larvæ, but nevertheless Professor Ewart * has been able to bring to a successful conclusion some experiments upon rabbits. Thus he found "that if a well-matured rabbit doe is prematurely (*i. e.*, some time before ovulation is due) fertilised by a buck of a different strain, the young take after the sire; when the fertilisation takes place at the usual time, some of the young resemble the buck, some the doe, whilst some present new characters or reproduce, more or less accurately, one or more of the ancestors. When, however, the mating is delayed for about thirty hours beyond the normal time, all the young, as a rule, resemble the doe. It may hence be inferred that in mammals, as in echinoderms, the characters of the offspring are related to the condition of the germ-cells at the moment of conjugation, the offspring resulting from the union of equally ripe germ-cells differing from the offspring developed from the conjugation of ripe and unripe germ-

* Presidential Address before the British Association. *Vide* Nature, vol. lxiv, p. 482, 1901.

cells, and still more from the union of fresh and over-ripe germ-cells." Upon plants, as far as I am aware, no direct experiments have been made, but some observations of De Vries * upon *Œnothera Lamarckiana* bear closely upon the matter. Some seeds of this plant had been kept for 5½ years before sowing, and it was then found that only about 1 per cent. of them germinated, instead of the usual 14 per cent. or so. Of the seedlings obtained, however, about 40 per cent. were "mutations," whilst the proportion of mutations obtained from fresh seeds was only 1 to 5 per cent.

Weismann's second factor in the production of variations is the so-called amphimixis, or sexual reproduction in multicellular organisms, and conjugation in unicellular. That this is one of the chief causes of variation was maintained by W. K. Brooks † some years ago. Basing his theory on Darwin's hypothesis of Pangenesis, he considered that as every "gemmule" of the spermatozoön united with that particle of the ovum which is destined to give rise in the offspring to the cell which corresponds to the one which produced the germ or gemmule, then such a cell will be a hybrid, and will therefore tend to vary. In his opinion the egg-cell is the conservative principle which controls the transmission of purely racial or specific characters, whereas the sperm cell is the progressive element which causes variation.

To what extent are we justified in assuming that this process of amphimixis does furnish an inexhaustible

* "Die Mutationstheorie," p. 360.

† "The Law of Heredity: A Study of the Cause of Variation and the Origin of Living Organisms," Baltimore, 1883.

supply of fresh combinations of individual variations, as Weismann maintains? Exact numerical evidence upon the point is, indeed, almost entirely wanting, but one's own everyday experience is all in favour of its validity. Thus one knows that animals of the same litter, which during embryonic development must have been exposed to very nearly equal environmental conditions, differ almost as much from each other as from animals of former litters, and in many cases not very much less than from animals in the litters of entirely different parents.

Now, as this phenomenon is one of almost universal occurrence, it cannot be maintained that the observed variations may be brought about by chance differences of environmental conditions acting during development. They must obviously be in chief part the result of differences in the individual sex-cells from which the offspring took their rise. So important is this conclusion that it was enunciated by Victor Hensen as a fundamental law of amphigonic heredity. This has been thus worded by Weismann:* "The individual is determined at the time of fertilisation, or, in other words, the individuality of an organism results from the fact that the germ-plasm is composed of the paternal and maternal ids which are brought together in the egg-cell."

Very interesting evidence in favour of this law is furnished by cases of identical human twins. It has long been known that whilst the larger number of twins show no greater resemblance to each other than do children of the same parents born consecutively, a

* "Germ-plasm," p. 253.

certain proportion exhibit a most striking resemblance, which, although not perfect, is much closer than has ever been observed in children born successively. These Weismann speaks of as "identical" twins. He says there is every reason to suppose that such twins are derived from a single ovum and spermatozoön, whilst dissimilar twins are derived from two ova, which must, of course, have been fertilised by two different spermatozoa. If this is actually the case, it furnishes a proof that heredity is potentially decided at the time of fertilisation.

Interesting cases of identical twins have been recorded by Galton in his book on "Inquiries into Human Faculty." With reference to disease, for instance, it was found that both twins were apt to sicken at the same time in 9 out of the 35 cases collected. Either their illnesses were non-contagious, or, if contagious, the twins caught them simultaneously. The mental and moral resemblance between the twins was just as close as the physical. An instance cited by Dr. J. Moreau * well illustrates this. His case was one of twin brothers who had been confined on account of monomania. They were physically so alike as to be easily mistaken for one another, and as regards their moral condition they had exactly the same dominant ideas; they both considered themselves subject to the same imaginary persecutions; they both had hallucinations of hearing; both were melancholy and morose.

Unfortunately, Galton did not obtain any exact anthropometric data. Weismann has obtained one series of measurements, however, viz., for twin brothers

* "Psychologie Morbide," Paris, 1859, p. 172.

seventeen years of age.* The following are the measurements made:

	TWIN A.	TWIN B.	PER CENT. DIFFERENCE.
Stature, ...	172 cm.	170 cm.	1.2
Left arm, ...	74	74	0.0
Right arm, ...	71	74	4.1
Left upper arm, ...	27	27.5	1.8
Fore arm, ...	27	26	3.8

These slight differences are probably due to the effect of external influences acting during the course of development, or are somatic, as distinguished from blastogenic, variations.

In a series of measurements on twin brothers obtained by the author, the resemblance was very much closer, the difference in no case reaching even 1 per cent. These brothers, aged twenty-three, were extraordinarily alike in physiognomy, and, moreover, they had both suffered at the same times from the same diseases, viz., bronchitis, measles, chicken-pox, mumps, and influenza. The slightly smaller one of the two had had a rather more severe attack of bronchitis than his brother, when a year and a half old, and so, perhaps, but for this, the physical resemblance would have been even closer:

	TWIN A.	TWIN B.	PER CENT. DIFFERENCE.
Standing height, ...	173.00 cm.	172.67 cm.	—.19
Sitting height (from seat of chair),	88.03	87.87	—.18
Span of arms, ...	179.90	179.88	—.01
Length of right mid-finger (from metacarpo-phalangeal joint),	10.99	10.98	—.09
Span of hand, ...	21.33	21.30	—.14
Length of skull (occipital protuberance to base of nose), . .	18.52	18.42	—.54
Maximum breadth of skull, . .	15.13	15.01	—.80

† "Germ-plasm," p. 253.

The finger prints, though bearing some resemblance, were nevertheless easily distinguishable. Hence in this case Galton's finger-print method would serve for an identification, whilst Bertillon's anthropometric system would be useless.

The very slight modifications produced in these twins by the action of environment during growth is probably explained by the fact that they had always been brought up together, and so exposed to practically the same conditions all their lives.

In another series of measurements upon twin brothers (aged twelve), the differences observed were somewhat greater, and the facial resemblance was likewise not quite so marked as in the previous case. The boys had both had scarlet fever, chicken-pox, and measles at the same times. Of the measurements given, it will be seen that the span of arms and length of forearm showed the greatest differences. As before, one twin was slightly smaller than the other in respect of every measurement made:

	TWIN A.	TWIN B.	PER CENT. DIFFERENCE.
Standing height, ...	143.22 cm.	142.62 cm.	− .42
Sitting height, ...	72.93	72.87	− .08
Span of arms, ...	150.73	148.81	−1.27
Elbow to tip of mid-finger, . . .	39.59	38.93	−1.67
Length of right mid-finger, . . .	10.06	10.03	− .30
Circumference of head (over occipital protuberance and 2.5 cm. above eyebrows), ...	50.23	49.97	− .52

From the vegetable kingdom Weismann also quotes an instance in support of Hensen's law. Thus, in respect of two species of *Oxalis*, "the flowers of the different hybrids were by no means quite similar, but three principal forms could be distinguished according to the combination of colours in the flowers. The flowers of the same hybrid, however, resembled each other in the most minute details. One plant, for instance, had violet petals of a rather pinker tint than those of one of the parent species, and all the petals were strongly tinged with red on one and the same lateral margin. As far as I could observe, all the flowers were similarly coloured on this stock. On another stock, all the sepals had brown rims, and on a third there was a narrow dark orange-coloured band in the centre of each flower. In these cases, therefore, the combination of the colours of the parents which appeared in the petals of the hybrids must have been decided at the time of fertilisation." *

That the influence which the maternal fluids can exert on an embryo during intra-uterine development is at best very slight seems at first sight to be proved by the experiments of Heape † on the transplantation of rabbits' ova. In the first successful experiment, two segmenting ova were obtained from an Angora doe rabbit which had been fertilised by an Angora buck thirty-two hours previously, and were immediately transferred to the upper end of the fallopian tube of a Belgian hare rabbit which had been fertilised three hours before by a buck of the same breed as herself. In due course this Belgian hare doe gave birth to six young. Four of these resembled herself and her mate, but the other two were undoubted Angoras. The Angora young were characterised by the possession of the long silky

* "Germ-plasm," p. 256.

† Proc. Roy. Soc., xlviii. p. 457, 1890.

hair peculiar to the breed, and were true albinoes, like their Angora parents. They also possessed the characteristic habit of slowly swaying the head from side to side when they looked at one. Both of the Angora young were born bigger and stronger than any of the other young, and they all along maintained their supremacy in this direction. Heape could observe no sign in the Angora young of any Belgian hare strain, and the Belgian hare young showed no likeness to their foster-brothers.

In a subsequent paper,* Heape records another successful experiment. In this a Belgian hare doe was covered by a Belgian hare buck, and shortly after the segmenting ova obtained from a Dutch doe which had been covered by a Dutch buck twenty-four hours previously were transferred to her fallopian tube. The Belgian hare doe gave birth to seven young, of which five were Belgian hares, and two very irregularly marked Dutch. It was found, however, on putting the same Dutch buck which had been used in this experiment to a thoroughbred Dutch doe, most, if not all, of the litter resulting were as badly marked as the Dutch foster-children. Hence it is not necessary to suppose that the foster-mother was the cause of the irregularity. From this and other evidence Heape considers he is justified in concluding that the uterine foster-mother exerts no modifying influence upon her foster-children, in so far as can be tested by the examination of a single generation. Romanes † has however remarked, that inasmuch as rabbits, when crossed in the ordinary way,

* Proc. Roy. Soc., lxii. p. 178, 1897.

† "Darwin and after Darwin," vol. ii. p. 146.

never throw intermediate characters, the result of Heape's experiment is without significance, as far as it bears on the inheritance of acquired characters. Heape considers that Romanes is mistaken in this view, for he has himself obtained experimental evidence to show that some of the young obtained by crossing are of an intermediate character. It is nevertheless true that in the majority of cases the young are apparently pure bred of one type or the other, and hence the value which ought to attach to Heape's experiments, so far as they relate to the production of somatic variations by change of environmental conditions during embryonic development, is probably not very great.

The evidence so far available seems to render it highly probable, therefore, that the major part of the variation exhibited by organisms is of blastogenic rather than somatic origin. It is due more to differences in the germ cells than to external influences acting during ontogeny. If it be found possible to collect considerable series of anthropometric measurements of identical twins, and to compare them, as regards variability, with similar measurements on dissimilar twins, then we may hope to obtain some adequate conception as to what proportion of the variation exhibited by adult individuals is due to external influences acting during pre- and post-natal existence. Also by comparing the variability of dissimilar twins with that of members of families produced in the normal manner of one at a birth, we may hope to determine what changes, if any, are produced by the slight differences in the maternal fluids which must doubtless exist during the development of the different offspring.

The variations of offspring are therefore largely produced by the mingling of dissimilar parental germplasms, so that the offspring do not closely resemble either each other or their parents. But there must clearly be a relation of some sort between them. As to the extent of this relation, we are chiefly indebted for our knowledge to the labours of Mr. Francis Galton. In his work on "Natural Inheritance," he has analysed a very large number of anthropometric data which were collected by himself specially for the purpose. The most important of them consist of records of the stature, eye-colour, artistic faculty, and condition of health of the various members of some 150 distinct families, extending over three or more generations. Arguing from these data, he concluded that on an average each parent contributed to the characters of his or her offspring $\frac{1}{4}$ of their amount, or both parents together contributed a half; whilst each grandparent contributed $\frac{1}{16}$, or the four grandparents together $\frac{1}{4}$, and so on; but he considered his data insufficient to warrant him in extending the sequence to more distant generations.

Some years later, Galton obtained other more favourable data.* These enabled him to ascertain the contributions of ancestors to offspring with much greater exactness, and warranted him in formulating a *Law of Ancestral Heredity*, which there is some reason for thinking may prove to be universally applicable to bisexual descent. The data consisted in long series of records of the colours of a large pedigree stock of Basset hounds, extending through many successive

*Proc. Roy. Soc., vol. lxi. p. 401, 1897.

generations. These records were preserved by Sir Everett Millais, who had originated the stock of hounds some twenty years ago. The Bassets are dwarf bloodhounds, of two and only two recognised varieties of colour. They are either white with large blotches ranging between red and yellow, or they may, in addition, be marked with more or less black. In the former case they are technically known as "lemon and white," and in the latter case as "tricolour." Transitional cases between these two forms are very rare. No less than 817 hounds of known colour, all descended from parents of known colour, were available as material. In 567 out of these 817 the colours of all four grandparents were known, and in 188 cases the colours of all eight great-grandparents were known as well. It was found that 79 per cent. of the parents of tricolour hounds were tricolour, whilst 56 per cent. of the parents of lemon and white hounds were tricolour. Hence from these values the contributions of unknown ancestors could easily be calculated. Working from these numerous data, Galton was able to confirm entirely his previous conclusions regarding heredity, and extend them in the direction then hinted at. He proved that the two parents do contribute between them one-half or (0.5) of the total heritage of the offspring: whilst the four grandparents contribute one-quarter, or $(0.5)^2$: the eight great-grandparents one-eighth, or $(0.5)^3$, and so on. Thus the sum of the ancestral contributions is expressed by the series [$(0.5) + (0.5)^2 + (0.5)^3 + \ldots$ etc.], which, being equal to 1, accounts for the whole heritage. The same statement may be put in a different form, by saying that each parent contributes on an average one-quarter, or $(0.5)^2$, each grandparent one-sixteenth, or $(0.5)^4$, and so on, or that the occupier of each ancestral place in the *n*th degree, whatever be the value of n, contributes $(0.5)^{2n}$ of the heritage.

It is unnecessary to quote the numerical details adduced by Galton, but two final results may be mentioned just to show how close was the approximation between fact and theory. Thus in one series 387 tricolour offspring were obtained from certain parents of known colour, themselves the offspring of parents of known colour. On the law of heredity, the number of tricolour offspring should have been 391. In the other series, the colours of the great-grandparents were known in addition, and in this case the approximation was even closer. One hundred and eighty-one tricolour offspring were obtained, as against the calculated number of 180.

Galton points out that there is nothing in his statistical law to contradict the generally accepted view that the chief, if not the sole, line of descent runs from germ to germ, and not from person to person. The person on the whole may be accepted as a fair representative of the germ, and so statistical laws which apply to persons would apply to germs also. Now the law is strictly consonant with the observed binary subdivisions of the germ cells, and the concomitant extrusion and loss of one-half of the several contributions from each of the two parents to the germ cell of the offspring.

Galton's law has been shown by Pearson * to be even

* Proc. Roy. Soc., lxii. p. 386.

more fundamental and far-reaching than its author claimed it to be. Thus he says, "If Mr. Galton's law can be firmly established, it is a complete solution, at any rate to a first approximation, of the whole problem of heredity." Professor Pearson points out that by means of it we are enabled to find the coefficients of correlation between an individual and any individual ancestor, and that these coefficients in their turn will suffice to determine all inheritance, whether direct or collateral.

As regards the relation of this law of heredity to variations produced by amphimixis, it is necessary to emphasize one fact, viz., that it concerns only the *average* contributions of ancestors to offspring, and not the absolute contributions. Within what limits the contributions of each parent and grandparent to the heritage of a child may vary, nothing whatsoever is known. It is possible that they may be very wide indeed, and everyday experience tends to give colour to this view. How trite is the expression that such and such a child is the "image" of his father or mother, whilst instances are no less common in which it is difficult, if not impossible, to trace any distinct resemblance between parent and child. Such cases as these, even if they could be substantiated by exact physical measurements, would in reality prove but little. It would be impossible to make accurate comparisons of all the tissues and organs of the body, and of the cells composing them, and it might be that these unexamined and unexaminable portions of the organism in reality possessed a very close correlation with the corresponding parental tissues. The average degree of correlation for all the tissues in the body might thus be just as great as for an individual who to all appearances closely resembled his parents, but in this case chiefly in external characters and not internal.

If amphimixis be so largely responsible for the variations observed in offspring, what, then, are the rules which govern the amount and range of these variations? For the answer to this question we are again primarily indebted to the labours of Mr. Galton. He set himself to determine the exact average relationships between the two parents and their offspring. It might be thought that this was so simple and obvious as to render it waste of time to put it to the test of experiment. It might be thought, in fact, that the average characters of offspring are a mean between those of the parents. But this is far from being the case. As Galton first showed, by means of extensive observations on the size of sweet-pea seeds obtained from plants which had been grown from seeds of known size, the average characters of the offspring show a considerable *regression* towards the mean characters of the race. That is to say, in the present instance, the size of the filial seeds was, on an average, more mediocre than that of the parent seeds. By means of his data, above referred to, concerning the stature of families, Galton was able not only to thoroughly substantiate this phenomenon of regression, but to calculate with some degree of exactness the actual amount of regression occurring between various kinsmen. For example, he found that if parents were sorted into groups according to their stature, then the stature of their sons, on an average, deviated only two-thirds as much from the mean stature of the general population as theirs did. Thus, if the mid-parental stature (the average between the statures of the man and the transmuted stature of the woman) be 72 inches, or $3\frac{3}{4}$ inches greater than the mean stature of the whole population, then the average stature of their sons will be only $3\frac{3}{4} \times \frac{2}{3} = 2\frac{1}{2}$ inches greater than the mean. If the mid-parental stature be 66 inches, or $2\frac{1}{4}$ inches less than the mean stature, then the average filial stature will be $66\frac{3}{4}$ inches, or only $1\frac{1}{2}$ inches less than the general mean stature. In addition to calculating the regression between parents and sons, and grandparents and grandsons, Galton calculated it for collateral relationships, as between uncles and nephews, and brothers and brothers.

Many of the data recorded by Galton in his "Natural Inheritance" were worked over again by Pearson in his memoir on "Regression, Heredity, and Panmixia,"* and various improvements in statistical methods suggested. The mathematical measure of regression, or *coefficient of regression*, he defined to be "the ratio of mean deviation of offspring of selected parents from the mean of all the offspring to the deviation of the selected parents from the mean of all the parents." It is to be noticed that according to this definition, the deviation of the offspring ought to be measured from the mean of the offspring of the general population, and not of the whole population, both parents and offspring, for thereby factors such as secular natural selection and reproductive selection are allowed for.

In this memoir, when discussing coefficients of re-

* Phil. Trans. 1896, A. p. 253.

gression, Pearson came to the conclusion that there was not at that time sufficient ground for forming any definite conclusion as to the manner in which lineal and collateral heredity were related. Thus it did not appear necessary to him that the coefficient of the former should be half that of the latter, as Galton had supposed. On attacking the problem a second time,* however, Pearson succeeded in proving that they were connected, according to a mathematically ascertainable relationship, so that, starting from Galton's law of heredity, it was possible to calculate the coefficients of regression or correlation between an individual and any of his kinsmen, either direct or collateral. Thus Pearson calculated the coefficient of regression between midparent and son to be .6, or somewhat less than that found by Galton. Between a single parent and a son, it would, therefore, be .3. Between grandparent and grandson it was .15, between great-grandparent and great-grandson .075, and so on. Between brothers it was .4, or considerably less than the coefficient found by Galton. Nevertheless this value confirms Galton's conclusion that brothers are more closely related to each other by blood than are fathers and sons.

It may be pointed out that in a stable population the coefficients of regression and of correlation between an individual and an ancestor are one and the same thing. If, however, the population is not stable, so that the variability of the offspring differs from the variability of the parents, then these coefficients also differ slightly.

The importance of this extension of Galton's law cannot be rated too highly, for by its means the whole

* Proc. Roy. Soc., lxii. p. 386.

theory of heredity is rendered simple, straightforward, and luminous. Pearson points out that we no longer need to know the characters of parents, grandparents, etc., to test the law, for any single relationship, near or far, direct or collateral, will bring its quota of evidence for or against it.

Galton's principle of "Regression towards Mediocrity" has been spoken of occasionally as if it were something abnormal and unexpected; something, indeed, unexplained and inexplicable. It is clearly nothing of the kind, however, but only what might readily be deduced from his law of Ancestral Heredity, supposing that this and this alone were known to us. Thus we have seen that offspring derive certain portions of their heritage from their grandparents and more remote ancestors, and as these are likely to be, on an average, more mediocre than their parents, they water down the parental characters transmitted to the offspring. Supposing all the grandparents and more remote ancestors of any given parents were absolutely mediocre, then, as the offspring receive only half their heritage from these parents, they would exhibit their characters in only half strength, or the coefficient of regression would be .5, and not .6. The reason why the regression reaches, on an average, the higher figure, is of course that the grandparents and other ancestors are *not*, as a rule, absolutely mediocre. They possess the characters exhibited by the parents, though in a diminished degree. Grandparents regress on parents to just the same extent as offspring do.

It may perhaps be enquired how it is that, if offspring are on an average more mediocre than their parents, the variability of the race does not become less and less for each generation, and so finally be reduced to zero. Why this is not the case is perhaps most readily grasped by examining a statistical table of the relations between parent and offspring in respect of some character. The table here given is reduced from a larger one given by Galton in his "Natural Inheritance" (p. 208), and represents the numbers of adult children of various heights born of 205 mid-parents of various heights. For instance, we see that the 17 mid-

NUMBER OF MID-PARENTS.	HEIGHT OF MID-PARENT IN INCHES	NUMBER OF ADULT CHILDREN OF HEIGHTS							MEDIAN HEIGHTS OF CHILDREN.
		below 63	63–65	65–67	67–69	69–71	71–73	73 & over	
5	73 & over						1	3	
17	71–73			4	8	18	22	10	70.6
63	69–71	1	18	23	62	90	42	15	69.1
82	67–69	4	37	92	131	126	37	3	67.9
32	65–67	4	22	37	49	29	3		67.0
6	below 65	3	14	9	8	3			65.6

parents 71 to 73 inches in height had between them 62 children, of whom the most frequently occurring were also 71 to 73 inches, or the same height as their parents. Children of from 69 to 71 inches were, however, nearly as frequent, so that, on the whole, it is obvious that the stature of the children was more mediocre than that of the parents. The median, or middle height of all these 62 values, was, in fact, only 70.6 inches, or 1.4 inch less than the median height of the mid-parents. The median of the children of mid-parents 69 to 71 inches in height, was .9 inch less than their median; whilst in the children of mid-parents varying from 67 to 69 inches it was only .1 inch less, for the median of these mid-parents, viz., 68 inches, was very nearly that of the whole population, and so obviously the filial height could undergo no regression, but would be practically the same value. The values in this table thus illustrate the existence of regression, but they also indicate that the offspring produced by these mid-parents are, as a whole, no less variable than they themselves are. The offspring are, in fact, more variable, as a mid-parental stature, being the mean of two parental statures, is obviously, on an average, less variable than either stature individually. Thus the mid-parents vary roughly between about 74 and 64 inches, but the children between 75 and 62 inches. This table therefore teaches us that though the children are, *on an average*, more mediocre than their parents, yet the general variability of the race is not diminished. The reason why the variability remains undiminished may be seen by studying the components of the vertical columns of the table. For example, with reference to children 71 to 73 inches in height, we see that mid-parents of 71 inches and upwards contribute proportionately more of these tall children than do any other mid-parents, but still mid-parents of 69 to 71 inches contribute (proportionately) a good many, and parents of 67 to 69 inches no small number. Even mid-parents of 65 to 67 inches contribute a very minute number of these children, who are thus no less than 6 inches taller than their parents. By these several contributions, therefore, the number of tall—and similarly of other—children is kept up to the same level in each generation. One may accordingly sum up the contents of this table as follows: Tall parents have many tall children, a moderate number of medium children, and a very small number of short children; medium parents have many medium children, and moderate numbers of tall and short children; short parents have many short children, a moderate number of medium children, and a very small number of tall children.

As was first pointed out by Mr. Galton,* characters such as stature and eye-colour offer a distinct contrast in their hereditary behaviour, for whilst "Parents of different statures usually transmit a blended heritage to their children, parents of different Eye-colours usually transmit an alternative heritage. . . If one parent has a light Eye-colour and the other a dark Eye-colour, some of the children will, as a rule, be light and the rest dark: they will seldom be medium eye-coloured." Thus eye-colour is a case of more or less exclusive inheritance, or inheritance by the offspring of the whole of the character of one parent and none of that of the other. Obviously, therefore, for such inheritance the law of ancestral heredity does not *at first sight* appear to hold. Supposing the offspring are equally likely to take after one parent or the other, then the coefficient of regression between parent and offspring will be .5, instead of .3, as in the case of blended inheritance: between grandparent and offspring it will be .25, instead of .15, and so on. Nevertheless it is probable that the law of ancestral heredity is just as true for one form of inheritance as for the other, only from the mere fact of the inheritance being exclusive, it does not reveal itself in the same way. Supposing there is no alternative between, for instance,

* "Natural Inheritance," p. 139.

light and dark eye-colour, or in animals, light and dark coat colour, then we can imagine the various light and dark heritages from each of the parents, grandparents, and more remote ancestors to be summated and balanced against each other in each individual, and, whichever reach the higher figure, be it by ever so little an amount, to be thereby enabled to originate exclusively the character to which they correspond. The constitution of the germ-plasm of a light or dark-coloured animal cannot be inferred, therefore, unless the colour of its ancestors be known, for it may contain anything from just over half right up to the full number possible of "light" or "dark" determinants.

That exclusive inheritance obeys the law of heredity in the same manner as blended inheritance seems to be shown by the fact that the striking proof of the law referred to above was obtained by Galton for an almost exclusively inherited character, viz., coat colour in Basset hounds. Galton believed also that his data for eye-colour in man afforded considerable support to the law in question. Professor Pearson,* however, seems to regard exclusive inheritance as distinct from blended inheritance, and to look upon it as governed by a *Law of Reversion*, and not by the law of ancestral heredity. Arguing from this law, we may suppose that 25 per cent. of the offspring show the full character of either parent, $\frac{25}{4}$ per cent. of them exhibit or revert to the full character of each of the four grandparents, $\frac{25}{16}$ per cent. revert to the full character of each of the eight great-grandparents, and so on. However, the whole

*Proc. Roy. Soc., lxvi. p. 140; *also* "Grammar of Science," pp. 486–496; *also* Phil. Trans. Roy. Soc. 1901, A. p. 79.

question is fraught with doubt and difficulty, and greatly lacks the experimental data necessary for putting it to an adequate practical test. Hence, for the present, it is best to regard the matter as still *sub judice.*

The law of heredity and regression which we may consider to have been substantiated for sweet peas, for Basset hounds, and for man, may justifiably be extended to other organisms as well. It seems probable that it is, in fact, to use Mr. Galton's words, "universally applicable to bisexual descent." As already stated, however, it should always be borne in mind that it deals only with average amounts, and not absolute amounts. Though the law is of great value in the breeding of pedigree stock, it is not exact enough to enable one to predict with any accuracy the characters of the unborn offspring of known parents, or even of known grandparents as well as parents. Nevertheless the mere fact of such a law of average inheritance being demonstrable, indicates triumphantly how fundamentally important is the constitution of the germ-plasm in the determination of variations.

We see, then, that Weismann's conclusions as to the chief factors concerned in the origin of blastogenic variations are in the main confirmed, so far as it is possible to put them to an experimental test. It is of course impossible to obtain experimental proof of the actual existence of biophors, determinants, and ids in the germ-plasm, but it is scarcely possible to account for the facts of heredity without making some such hypothesis. The Law of Ancestral Heredity proves that all ancestors, however remote, are able to leave the impress of their individuality upon the sex-cells, in diminishing proportion according to their remoteness. Such a fact can only be accounted for by assuming the existence, in the germ-plasm, of definite units carrying definite characters, and the regular halving in the average strength or amount of such characters during the reducing division of the nuclear matter of the sex-cells which precedes each act of sexual reproduction.

It was stated above that Professor Pearson calculated the correlation constant between brothers to be .4. In a remarkable memoir recently published Professor Pearson * and his collaborators have collected together all the statistics at present available, as to fraternal correlation in the horse, the dog, and in daphnia, as well as in man, and have found the mean of the constants deduced from 19 series of observations to be .4479. Individual constants range from .6934 down to .1973, but doubtless some of the extreme values in either direction are, for various reasons, invalid. It seems probable, therefore, that fraternal correlation, whether it concerns stature, cephalic index, eye-colour, or longevity in man, or coat colour in the dog and horse, may be taken to fluctuate about a mean value of .4 to .5.

The greater part of this memoir, however, concerns correlation in the vegetable kingdom. Professor Pearson points out that the individual puts forth a number of like organs, such as blood corpuscles, spermatozoa, petals of the flower, leaves of the trees, which are undifferentiated, but that nevertheless there is a considerable amount of variation among these "undifferenti-

* Phil. Trans. 1901, A. p. 285.

ated like organs," or "*homotypes.*" It is found, however, that the variability of these like organs in an individual is less than that of similar like organs in all the members of a race (it being as a rule 80 to 90 per cent. as great), and that therefore there is a considerable correlation between them. The principle that like organs are correlated, or that the undifferentiated like organs of individuals have a certain degree of resemblance, Professor Pearson speaks of as *homotyposis.*

Professor Pearson and his collaborators have determined the degree of homotyposis in 22 distinct series, and have determined, for instance, the numbers of leaflets on the leaf of the Ash (26 leaves being taken from each of 329 trees), the number of veins in the leaf of the Spanish Chestnut (26 leaves from 204 trees), and in that of the Beech (26 leaves from 100 trees), the prickles on the leaf of the Holly (26 leaves from 299 trees), the stigmatic bands on the seed capsules of poppies (10,435 capsules, taken from 1064 plants), the *sori* on 8 to 12 fronds of each of 101 Hartstongue ferns, etc. The mean correlation for all the 22 series was .4570, or practically the same value as was obtained for fraternal correlation. The extreme values ranged from .6311 to .1733, but there are numerous causes which will account, at least in part, for these wide deviations from the average. Supposing that any of the organs measured had undergone a certain amount of differentiation or splitting up in various directions (and this, it must be remembered, is always possible, as there is no real criterion as to whether any given organ is really undifferentiated, or differentiated), this would generally result in a great reduction in the correlation;

whilst a heterogeneity of material, such as a mixture of two different local races, would tend, as a rule, to raise correlation.* Also the environmental factor, and the difficulty of ensuring that all individuals are of the same age, or in the same state of development, must be borne in mind.

Professor Pearson therefore considers that he is justified in assuming that the intensity of pure homotyposis throughout the vegetable kingdom probably lies between .4 and .5, and as this is the mean value for fraternal correlation, he believes that "heredity is really only a phase of the wider factor of homotyposis."

In a criticism of Pearson's conclusions, Bateson † draws attention to the fact that it is difficult or impossible to distinguish between chance variation occurring between members of a series, and actual differentiation, which may be present in greater or less degree. He therefore considers that the average value of the homotyposis coefficient has no significance. However, Pearson states that the "diversity due to differentiation. . . is the result of dominating factors which can be isolated and described," though he does not attempt this in detail in his present memoir. To what extent he will be able to accomplish it, and so ultimately obtain the true correlation constants of absolutely undifferentiated like organs, remains to be seen.‡

* *L. c.*, p. 292.

† Proc. Roy. Soc., lxix. p. 193.

‡ See also rejoinder by Professor Pearson in Biometrika, i. p. 320, 1902.

CHAPTER V.

BLASTOGENIC VARIATIONS (*Continued*).

Reversion; commonest in crossed races, as of the pigeon and fowl; its theoretical explanation—Prepotency; in the trotting horse and in man; probably due in large part to inbreeding—Mendel's Law of Hybridisation, and its range—Natural and artificial plant hybrids—Animal hybrids—Sports; probably of different origin to normal variations—Artificial production of monsters—Telegony; probably non-existent—Parthenogenesis in an Ostracod and in Daphnia—Does sexual reproduction induce variability?—Relation of variability of individual to variability of race—Asexual reproduction in plants—Bud-variation.

In the last chapter we saw that the average characters of offspring are inherited from their ancestors in accordance with a simple and definite law, but it remains for us to discuss several phenomena related to this law, some of which, indeed, appear to afford a partial contradiction of it. These are the phenomena of reversion, prepotency, the appearance of sports, and certain cases of hybridism. The variations which show themselves in connection with such phenomena, though doubtless of less importance than those already discussed, are nevertheless in some instances considerable, and of not infrequent occurrence. They therefore merit a fairly full discussion. It is impossible, however, to illustrate this with many exact numerical data, simply because these do not exist. One must as a rule remain content to quote the descriptive evidence of breeders and others, who seldom troubled to substantiate their views by measurements and figures.

The phenomenon of *Reversion* or *Atavism* has long been recognised, not only by agriculturalists and breeders, but also by others who have witnessed its occurrence in members of the human race. One of the simplest instances of reversion is that of a child or a lower animal resembling a grandparent more closely than its immediate parents. Much more remarkable, however, are those instances in which the resemblance is to a remote ancestor, or to some distant member in a collateral line (supposing, of course, that these be held to be properly substantiated). Cases of reversion are very frequent in respect of secondary sexual characters, as when a son resembles his maternal grandsire more closely than his paternal in some such attribute as a peculiarity of the beard, in the case of man; of the horns, in the case of the bull; and of the hackles or comb in the cock. Also it is well known that certain diseases, such as hæmophilia and colour-blindness, are frequently transmitted to male offspring through a woman who herself remains unaffected.

For most of our knowledge on the subject of reversion we are indebted to the labours of Charles Darwin, who obtained most valuable experimental evidence himself, besides collecting from most varied sources the results obtained by others. One of the most striking instances he records is that of a pointer bitch,* which produced seven puppies. Four of these were marked with blue and white, which is so unusual a colour with pointers that the animal was thought to have played

* "Animals and Plants," ii. p. 8.

false with the greyhounds, and all but one of the litter were destroyed. Two years later, this young dog was seen by a friend of the owner, and he declared him to be the image of his old pointer bitch, the only blue and white pointer of pure descent he had ever seen. On close enquiry, it was proved that the dog was the great-great- grandson of the bitch, and so, on Galton's law, it should have received only $\frac{1}{256}$ part of its heritage from her. Another even more remarkable instance is that of a calf which was coloured in a very peculiar manner, its legs, belly, and part of the tail being white, and the remainder black. Its great-great-great-great-grand-father was coloured in the same peculiar manner, but all the intermediate offspring were black. Hence the calf reverted in its colour markings to an ancestor from which it should have drawn only $\frac{1}{4096}$ part of its heritage.

It is when two distinct races are crossed that the tendency in the offspring to reversion most often declares itself. No examples are more striking than those obtained by Darwin in the case of the domestic pigeon. For instance,* he paired a mongrel female Barb-fantail with a mongrel male Barb-spot, neither of these mongrels having the least blue about them. "Nevertheless the offspring from these two mongrels was of exactly the same blue tint as that of the wild rock-pigeon from the Shetland Islands over the whole back and wings; the double black wing bars were equally conspicuous; the tail was exactly alike in all its characters, and the croup was pure white; the head, however, was tinted with a shade of red, evidently de-

* *L. c.*, i. p. 209.

rived from the Spot, and was of a paler blue than in the rock-pigeon, as was the stomach. So that two black Barbs, a red Spot, and a white Fantail, as the four purely-bred grandparents, produced a bird exhibiting the general blue colour, together with every characteristic mark, the wild *Columba livia.*"

Professor J. C. Ewart, in the breeding experiments he has recently been carrying out at Penycuik,* has obtained an equally striking case of reversion in the case of the pigeon. He crossed a pure white Fantail cock with the offspring of an Owl and an Archangel. One of the young of this complex pair had the colouration of the Shetland rock-pigeon, whilst the other resembled the Indian rock-pigeon in having a blue croup and the front part of the wings chequered. In this second bird there was complete reversion as to colour, and in the first, wherever measurements were possible, there was practically complete reversion also as to form. The tail feathers were twelve in number and showed but the faintest indications of any colour inheritance from their immediate parents. An additional point of interest lay in the fact that in disposition the bird seemed wilder and more shy than the domestic breeds usually are.

Many other instances might be quoted from Darwin and others to prove that this tendency to the production of offspring of a blue colour, with the same characteristic marks as *Columba livia*, is present in all the chief domestic races of pigeon. It shows itself more especially when these domestic races are crossed, but may even appear occasionally in the purely bred races.

* "The Penycuik Experiments," Edinburgh, 1899.

Similar phenomena show themselves in other domestic animals besides the pigeon, though they are seldom so striking or so clear. Thus in some cases the wild ancestor or ancestors are quite unknown, and hence one is debarred from coming to any certain conclusions as to whether reversion is present or not. The Game fowl, however, and probably most other domestic breeds of fowl, may with considerable confidence be derived from the jungle fowl, *Gallus bankiva.* Now purely bred Game, Malay, Cochin, Dorking, Bantam, and Silk fowls may frequently or occasionally be met with, which are almost identical in plumage with the wild *Gallus bankiva.* The most striking instance obtained by Darwin * is one in which a glossy green-black Spanish cock was crossed with a diminutive white Silk hen. Both of these breeds are ancient, and have long been known to breed true. All the offspring from this cross were coal black, and all plainly showed their parentage in having blackish combs and bones; but as the young cocks grew, one became a gorgeous bird, closely resembling the wild *G. bankiva*, but with the red feathers rather darker. In all but a few details there was the closest resemblance.

In recent years a series of interesting observations has been carried out by von Guiata upon mice.† Fifty-five Japanese waltzing mice were crossed with white mice belonging to a race bred by Weismann for eight years, and these crosses were continued through

* *L. c.* ii. p. 253.

† Ber. Naturf. Ges. zu Freiburg, Bd. x. p. 317, 1898, and Bd. xi. p. 131, 1900. Review by Davenport in Biol. Bulletin, ii. p. 121, 1901.

seven generations. Japanese waltzing mice are mostly black and white, *i. e.*, piebald, in colour, but their crosses and reciprocal crosses with the albino race yielded a most unexpected result. The whole of the offspring produced were of a gray colour, indistinguishable in respect either of colour or of size from the common house mouse. The waltzing action was entirely wanting, the reversion being apparently complete. Heacke had obtained a similar result on crossing the same races.* In the third generation, however, the type was broken, for the 44 young produced by the mating of 4 pairs of the reverting gray mice consisted of 8 waltzers (albino, spotted, gray, and black), 11 pure albinos, and 25 gray mice. In the subsequent generations, the albinos and also the gray and the spotted mice were found to breed true. Gray mice crossed with white yielded mostly gray offspring, but a certain number of waltzers.

Of the occurrence of reversion there can thus be no question. In fact, its appearance in the offspring of crossed races is by no means an infrequent phenomenon. The reversion of hybrids and mongrels to one of their pure parent forms, after an interval of two or more generations, is especially common. Hence it would seem that the act of crossing in itself gives an impulse towards reversion. Why and how this is the case must be more or less a matter of conjecture. Indeed, this is equally true for all the phenomena of reversion, but I think that a brief consideration of certain presumptions regarding the germ-plasm as the bearer of hereditary characters will show that, after all, we are not dealing

* Biol. Central., Bd. xv. p. 44, 1895.

with anything more mysterious and remarkable than is found in most of the phenomena of nature. Thus taking it for granted that each of the parts of an organism capable of independent variation from the germ onwards has a definite representative or determinant of some sort in the germ-plasm, what proportion does the mass of the determinants of, say, all the characters which distinguish a pouter or a fantail pigeon from a rock pigeon, bear to the mass of the determinants which represent the species pigeon, as such? Let us suppose that the average total differences between the characters of species of the same genus be counted as one unit, what would be the number of units corresponding to differences between the characters of genera, families, orders, and so on? No two biologists would judge alike, and of course it is impossible to estimate them; but, for the sake of our argument, let us attempt some sort of rough numerical estimate as to what these differences might be. Let us assume that, if species on an average differ by one unit in the sum total of characters, genera differ by three units, and families by perhaps ten units. Orders might differ by 25 units, classes by 50 units, and phyla by 100 units. Therefore we assume that an individual of one phylum, in the sum total of its characters, is 100 times more different from an individual of another phylum than is one species from another of the same genus. The difference between the highest Vertebrate and the lowest Protophyte would probably be considered to be perhaps ten times greater than this, but let that pass. Let us take it that the sum total of characters represented by any species of pigeon is 100 units, of which the total characters peculiar to a rock or other pigeon, as such, is one unit. Also let it be granted that the characters separating any variety of the pigeon from the ancestral rock pigeon are of the same value as those separating species of the same genus, namely, one unit. Now in the gradual course of evolution of a domestic variety of pigeon from a rock pigeon, we may assume that the total amount of germ-plasm bearing hereditary characters has remained practically constant, and hence, as one unit of determinants has been added on to the rock pigeon germ-plasm, one must have disappeared. Now did this unit of determinants corresponding to the characters of the domestic variety of pigeon replace that of the rock pigeon, or was it superimposed on it? Embryology seems to teach us that once any character is, as it were, laid down in the germ-plasm, it is fixed there, and as a rule only very slowly dwindles away by a process of gradual dilution by subsequent ontogenetic stages. It seems reasonable to assume, therefore, that the determinants are replaced in proportion to the relative amounts of them present, or that, on an average, $\frac{99}{100}$ of the replaced unit concern the sum total of hereditary characters which go to constitute the species *pigeon*, and $\frac{1}{100}$ those peculiar to the species *blue rock pigeon*. The germ-plasm of a domestic pigeon will therefore be made up of 1 unit of determinants corresponding to the characters *domestic pigeon*, .99 of a unit corresponding to the characters *blue rock pigeon*, and 98.01 units corresponding to the characters *species pigeon*. It therefore follows that the hereditary characters of the ancestral rock pigeon are almost as strongly represented in the germ-plasm of a domestic pigeon as they were originally, only that they seldom have an opportunity of showing themselves. They are covered up by the more recently acquired characters, and it is only under exceptional circumstances that they are able to reveal themselves. When, for instance, two distinct races of pigeon, such as a pouter and a fantail, are crossed, then the offspring would on an average receive .5 of a unit of determinants corresponding to each of the special group of characters *pouter* and *fantail*, the same .99 of a unit corresponding to the characters *blue rock pigeon*, and 98.01 units corresponding to the characters *species pigeon*. If, then, the determinants of pouter and fantail do not to any great extent correspond, what wonder is it that they more or less neutralise each other, and allow the *blue rock pigeon* determinants to gain the upper hand, and show their presence?

This view of the constitution of the germ-plasm may at first sight seem contrary to the law of ancestral heredity, but in reality it is not so. A man may receive a quarter of his hereditary characters from each parent, and a sixteenth from each grandparent, but all except a very minute proportion of these characters are common to all men, they being, in fact, the characters proper to the species *Homo sapiens*, as such. Instead of a quarter of a unit from each parent, a man in reality receives only a hundredth or a thousandth of a unit of characters peculiar to the parent as such, all the rest being the characters common to all members of the race. Even this minute fraction of a unit does not in any way represent characters acquired by the parent during his life-time, but is itself built up of proportions of peculiar characters received from his parents, grandparents and other ancestors in accordance with the law of heredity.

It seems, then, that the sudden reappearance of ancestral characters ought not to be regarded as a very remarkable phenomenon, but certain other cases of reversion offer a greater difficulty. Thus cases such as that above mentioned of a calf reverting to the colour marking of an ancestor six generations back, if of at all frequent occurrence, are truly remarkable. If of only very infrequent occurrence, however, they may perhaps be ascribed to a mere coincidence, or to like conditions of environment having acted on both ancestor and descendant, and produced like results.

Prepotency. The phenomenon of prepotency of certain individuals, races, and species in the transmission of their characters is a very common one, and it merits our consideration, in that it is an important factor in the production of variations. As a rule, the offspring of dissimilar parents are in most respects of an intermediate character. Frequently, however, they more or less closely resemble one parent in one part, and the other parent in another part. Less seldom they show a much closer resemblance to one parent than to the other, or may apparently resemble one parent in every respect, to the entire exclusion of the other parent. We may here be dealing with true cases of prepotency, or it may be that the characters in question are for some unknown reason unable to blend, and so be necessarily transmissible only *in toto* from one parent to the other. For instance, it is well known that certain domestic animals, such as the cat, show only a few well-defined differences of colour marking, such as white, black, tabby, and tortoise-shell, and though they breed promiscuously, very seldom throw intermediate colours. Again, Sir R. Heron crossed during many years white, black, brown, and fawn-coloured rabbits, and never once got these colours mingled in the same animal, but often got all four colours in the same litter.* All the offspring of dissimilarly coloured parents may therefore resemble either parent, or some resemble one and others the other, possibly quite apart from any question of prepotency.

Of undoubted cases of prepotency, but few have been recorded with much detail or exactness. Of those collected by Darwin, the most striking is that of a famous black grayhound,† which "invariably got all his puppies black, no matter what was the colour of the bitch"; but this dog "had a preponderance of black in his blood both on the sire and dam side," a point which will be referred to again later. Again, the famous bull Favourite is believed to have had a prepotent influence on the shorthorn race. The male Manx cat appears to be prepotent in transmitting his tailless condition. Professor Ewart‡ has recorded a few additional cases of prepotency. Thus a well-known breeder of highly bred ponies used to boast that he had a filly so prepotent through inbreeding that, though she were sent to the best Clydesdale stallion in Scotland, she would throw a colt showing no cart-horse blood, provided always that the Clydesdale was not also the product of inbreeding. Again, Professor Ewart points out that Jews, as a race,

* "Animals and Plants," ii. p. 70.

† *L. c.*, ii. p. 40.

‡ "Penycuik Experiments," p. xli.

are strongly prepotent, probably because they are purer bred than other races.

A numerical estimate of the frequency with which different grades of prepotency are distributed appears to have been attempted for the first time quite recently by Mr. Galton.* From data given in Wallace's Year Book of American Trotting Horses, he has determined the numbers of offspring of a certain standard, produced by various sires and dams. A standard performer is a horse which has succeeded in trotting a mile in 2 min. 30 seconds or less, or in pacing (ambling) a mile in 2 min. 25 seconds or less. Data concerning the offspring of 716 sires and 494 dams were available, and the following were the percentage proportions of "standard performers" produced by them.

	NUMBER OF STANDARD PERFORMERS PRODUCED BY A SINGLE PARENT, SIRE OR DAM.											Total Parents
	1	2	3	4	5	6 to 10	11 to 20	21 to 30	31 to 40	41 to 50	51 and above	
Sires	46	17	10	7	3	9	4	1	1	1	1	100
Dams	50	35	10	3	1	1	–	–	–	–	–	100

This table would seem to show that the prepotency of certain sires is enormous, even allowing for the tendency of breeders to send the best mares to the best horses. Thus the horse *Happy Medium* had 92 distinguished offspring, and *Electioneer* no less than 154. The same results are indicated by the produce of the dams, though the figures are less striking owing to the relative fewness of their offspring. A sire produces

* Nature, vol. lviii. p. 246, 1898.

30 foals annually, but a dam only one, hence the production of respectively 7, 8, and 9 standard performers by three mares is very remarkable. Professor Pearson,* however, does not accept the high degree of prepotency which these figures seem to indicate. He points out that the fact of certain sires producing such a preponderance of standard performers is largely due to their exceptional pedigrees. It is also due to the second-rate stallions being given far less chance of producing performers, in that the mares sent them are often inferior, or past their most intense fecundity, as well as being fewer in number.

In discussing the law of heredity in the last chapter, it was tacitly assumed that the heritage from each parent was the same, or that both parents were equipotent. This does not seem to be necessarily the case, however, as Professor Pearson finds that in man the father is slightly prepotent over the mother for the offspring of both sexes.† Thus a determination of the coefficient of correlation in respect of stature and of head index, gave the following figures:

		Coeff. of correlation
Father and son	(Middle class English)	.396 (for stature)
" " daughter	" " "	.360 " "
Mother and son	" " "	.302 " "
" " daughter	" " "	.284 " "
" " son	(N. American Indians)	.370 (for head index)
" " daughter	" "	.300 " " "

The average correlation between stature of father and of offspring was thus .378, and between that of mother and of offspring .293, or 22.5 per cent. less. The

* Nature, vol. lviii. p. 292.

† "Grammar of Science," p. 458.

average correlation between parent and offspring was thus .335, instead of the theoretical .3. Pearson thinks this high value may be due to assortive mating.* The number of data available for calculating these constants was not very great, so that they cannot be accepted as final, but there seems little doubt of the existence of a small degree of male prepotency. It should be noticed, also, that the intensity of the heredity is stronger in the son than in the daughter, and this not only for stature in the English race, but also for cephalic index in the North American Indians. A similar relation was found in respect of eye-colour, hence Pearson considers that in man it may be a general rule for the male to inherit more than the female.

A comparison of other coefficients of correlation seemed to show that the hereditary resemblance between brother and brother, or sister and sister, is greater than that between brother and sister. This was true for stature, head index, and eye-colour in man, and also for coat-colour in thoroughbred-race horses. It would therefore follow that inheritance in a line through one sex is prepotent over inheritance with a change of sex, or that, for instance, a man would resemble his paternal more closely than his maternal grandfather.

It is not to be imagined that prepotency of the male over the female is in any way a general law. Thus in thoroughbred horses, sire and dam are equipotent in the transmission of coat-colour. In Basset hounds, on the other hand, Galton † found that the female was prepotent over the male in transmitting colour in about the

* *L. c.*, p. 457.

† Proc. Roy. Soc., lxi. p. 404, 1897.

proportion of 6 to 5. Whatever the degree of prepotency of one parent over another—and with similarly bred stocks it is probably never very great—we must conclude that the average contribution of both parents together still remains at a half, that of grandparents at a quarter, and so on. It is only the relative proportions contributed by the two sexes which differ. Thus it will be remembered that it was the Basset hound data which afforded Galton such valuable evidence in support of his law.

But what view are we to take of the more striking instances of prepotency mentioned above? Are they also conformable to the law of heredity, or are they abnormal and exceptional? Galton himself has come to the conclusion that high prepotency does not arise through normal variation, but must rank as a highly heritable sport. As has been mentioned in a previous chapter, there is no adequate proof that sports transmit their characters more persistently than other variations, and in any case it is probably unnecessary to assume that prepotency is other than a special case of the law of heredity. Thus we saw in the above-mentioned case of the black grayhound, that the dog had a preponderance of black in his blood, both on the sire and dam side, whilst both the instances of prepotency mentioned by Professor Ewart seem largely attributable to inbreeding. This inbreeding, according to Professor Ewart, induces prepotency by fixing the characters of the particular variety selected. But what is really meant by fixing a character? To adequately comprehend the meaning of the term it is only necessary to examine Galton's law of ancestral heredity a little more in detail than we did in the last chapter, and it will then be obvious that a character becomes more and more fixed in the offspring, the more and more fully it is represented in the parents, grandparents, and more remote ancestors. Thus, according to the law, offspring receive a half of their heritage from their parents, a quarter from their grandparents, and so forth. But of what is this half and this quarter made up? Obviously half of the parental half heritage, or a quarter in all, was received from *their* parents, and a quarter of a half, or an eighth in all, from their grandparents, and so on; whilst, as regards the quarter heritage received by the offspring from their grandparents, a half of it, or an eighth in all, was received from *their* parents, and so on. By tracing back the heritages in this way, it is therefore possible to calculate the absolute amounts of any character or strain present in offspring, as distinguished from the relative amounts; relative, that is, to those present in the parents and other ancestors. For instance, supposing the parents and parents only had been selected in respect of any particular character, the condition of the previous ancestors being entirely unknown, then, as we have already seen, the offspring will exhibit these exceptional characters in only .6 of their full strength, or will have regressed to this extent towards the general mean of the race. Supposing the grandparents have been selected in respect of the same characters, as well as the parents, then Pearson * has calculated that the offspring will exhibit the characters in .8049 of their full strength; if the great-grandparents also, then in .9027 of their strength; and if still

* Proc. Roy. Soc., lxii. p. 399, 1898.

three other generations back be selected, then in .9879 of their full strength. That is to say, " after six generations of selection the selected individuals will, without further selection, breed true to the selected type within nearly 1 per cent. of its value." In fact, practically all regression towards mediocrity will have been weeded out. Supposing now, some variety of a species which had been bred true to its varietal characters for only two generations were crossed with another variety of the same species which had been bred true to *its* characters for six, then the resulting offspring would receive $\frac{.8049}{2} = .4025$ of the characters of one parent, $\frac{.9879}{2} = .4940$ of those of the more thoroughbred parent, and .1035 of unknown blood. Knowing as we do that many characters show little or no tendency to blend, it would not be very remarkable if the offspring resembled the more thoroughbred parent to the partial or almost complete exclusion of the ill bred. That is to say, the one parent would prove itself strongly prepotent, simply through its characters having become fixed through inbreeding.

Hybridisation. Though, as a rule, intercrossing between different varieties of the same species tends to produce uniformity of character, yet it may also very frequently lead to the production of increased variability, not only by the partial or complete absence of blending of the parental characters, but also by the appearance of seemingly fresh characters, due to reversion or some other cause.

Our knowledge of the laws governing hybridisation

is chiefly derived from observations on plants, for by reason of the ease and success with which they are carried out, and the scientific and practical results obtained, these altogether outweigh the comparatively few observations which have been made on animals. Though much of the evidence obtained is variable and contradictory, yet some of it has afforded results of striking lucidity. Especially is this the case as regards what may be termed Mendel's Law of Hybridisation. Though discovered as long ago as 1865,* this important generalisation has passed almost unnoticed until the last year or two, when it was independently re-discovered and confirmed by De Vries, by Correns, and by Tschermak. Mendel's observations extended over eight years, during which over 10,000 plants were examined. Most of them were made upon the different varieties of the pea, *Pisum sativum*. The varieties employed differed in respect of (1) the form of the ripe seeds, these being either nearly round, or angular and wrinkled; (2) the colour of the cotyledons, these being various shades of yellow or green; (3) the colour of the seed coat, this being either white, gray, or brown; (4) the form of the ripe pods, these being either simply inflated, or deeply constricted between the seeds; (5) the colour of the unripe pods, this being yellow, or light to dark green; (6) the position of the flowers, either axial or terminal; (7) the length of stem.

On uniting each of these two differentiating characters by cross-fertilisation, the hybrids obtained in each case were found to resemble only *one* of their

* Abhandl. d. naturforsch. Ver. in Brünn, Bd. iv. 1865. Translated by W. Bateson in J. Roy. Horticult. Soc., xxvi. p. 1, 1901.

parent forms, and to show little or no trace of the other. The characters thus appearing were termed by Mendel *dominant*, and the characters becoming latent in the process, *recessive*. In the next generation, however, the seeds from these dominant hybrids betrayed their mixed origin, for instead of maintaining the pure dominant character, on an average one out of every four of the plants or seeds obtained reverted to the *recessive* parent form. The following are the actual numbers of plants and seeds examined by Mendel in respect of the various differentiating characters above mentioned:

		Proportion of Dominant to Recessive.
(1)	253 hybrids (gave 7324 seeds)	2.96 : 1
(2)	258 " (" 8023 ")	3.01 : 1
(3)	925 "	3.15 : 1
(4)	1187 "	2.95 : 1
(5)	580 "	2.82 : 1
(6)	858 "	3.14 : 1
(7)	1064 "	2.84 : 1
		Mean 2.98 : 1

It will be seen that the proportion of 3:1 is fairly evenly maintained in respect of all the characters observed.

The observations on the next and succeeding generations afforded an even more remarkable result than this, for they proved that the recessive forms obtained in the second generation were absolutely pure. Thus the seeds obtained by crossing them amongst each other, or by self-fertilisation, yielded offspring which never showed any trace of the dominant grand-parental characters. The dominant forms, on the other hand, which of course were self-fertilised, underwent a further splitting up. A third of them yielded plants which in subsequent generations proved themselves to be pure dominant forms, whilst two thirds of them still retained their hybrid nature, as was shown by their yielding, in the next generation, recessive and dominant forms in the proportion of 1:3. The gradual resolution of the original hybrid forms into pure parental forms may be represented diagrammatically thus,* it being assumed that 64 hybrids with yellow cotyledons had been produced by the crossing of parental forms having respectively green and yellow cotyledons.

Parents	I. Gen.	II. Gen.	III. Gen.	IV. Gen.	V. Gen.
green		16 green	16 green	16 green	16 green
				8 green	8 green
			8 green	4 green	4 green
					2 green
	64 yellow	32 yellow	16 yellow	12 { 8 yellow	4 yellow (hy.)
		48	24		2 yellow
			8 yellow	4 yellow	4 yellow
				8 yellow	8 yellow
yellow		16 yellow	16 yellow	16 yellow	16 yellow

We see that of the plants produced by crossing the original 64 yellow hybrids haphazard amongst themselves, a quarter are of the pure green form, a quarter of the pure yellow form, and a half of them hybrids with the yellow character dominating. On crossing these hybrids among themselves, we see that in each subsequent generation their number is reduced by half, till in the seventh generation only 1 of the original 64 hybrids would be still remaining.

The explanation of this result was clearly laid down by Mendel, he supposing that the cross-bred plant produced pollen grains and ovules, each of which bore only *one* of the alternative varietal characters, and not both. If D and R represent the two characters present in dif-

* Modified from Correns (Ber. d. deutsch. bot. Gesell., xvii. p. 162, 1900.

ferent ovules of the hybrids, and d and r those in pollen grains, then on crossing these hybrids haphazard, the germ cells giving rise to the next generation will unite so as to form $Dd + Dr + dR + Rr$. Now Mendel found that it was perfectly immaterial whether the dominant character belonged to the male or the female plant, and so it follows that we should get twice as many similar hybrid forms (Dr and dR) as of pure dominant or pure recessive.

If parental forms possessing two or more differentiating characters be crossed, the law of alternative heritage continues to hold, though it necessarily becomes somewhat more complicated. For instance, Mendel crossed seed parents with round seeds (A), and yellow cotyledons (B), with pollen from plants having angular seeds (a), and green cotyledons (b). The hybrids would therefore consist of plants with germ cells having the characters AB, Ab, Ba, and ab. These hybrids, on crossing haphazard, would yield the following:

$$(AB + Ab + Ba + ab)^2 = \underset{38}{A^2B^2} + \underset{35}{A^2b^2} + \underset{65}{2A^2Bb} + \underset{60}{2AB^2a} +$$

$$\underset{138}{4\,ABab} + \underset{67}{2\,Aab^2} + \underset{68}{2\,Ba^2b} + \underset{28}{B^2a^2} + \underset{30}{a^2b^2}$$

The figures underneath indicate the actual numbers of plants obtained by Mendel from the 556 seeds yielded by the 15 original hybrid plants. The average numbers with two, three, and four characters are respectively 34, 65, and 138, or very nearly in the theoretical proportion of 1:2:4.

Mendel even took the immense trouble to cross parents differing in respect of three characters, and he found that the offspring of the resulting hybrids with

3, 4, 5, and 6 characters were on an average respectively as 10 : 19 : 43 : 78, or very nearly as 1: 2: 4: 8. He also confirmed his law by some observations on *Phaseolus vulgaris* and *P. nanus*, but the crossings of *P. nanus* ♀ with *P. multiflorus* gave only a partial result, whilst those on *Hieracium* did not agree at all. In the light of the hybridisation experiments of Kölreuter, Gärtner, and others, Mendel recognised that his law was by no means universally applicable. It obviously can only apply to cases of exclusive inheritance, and not to those of blended or mixed inheritance.

De Vries * has made similar observations to those of Mendel upon varieties of no less than 15 different species of plants, and in every case found that the proportion of recessive forms obtained in the second generation was approximately the theoretical 25 per cent. When the observations were continued through other generations, the results likewise agreed with theory. Tschermak † repeated Mendel's observations upon the different varieties of *Pisum sativum*, and with some of them obtained a similar result. However, he found that in some other cases ‡ the law did not hold. Correns § also experimented with varieties of the pea, and he found that whilst some of the characters obeyed Mendel's law, others, such as the colour of the skin of the seed, did not. He obtained a similar result ¶ on crossing *Matthiola incana*

* Ber. d. deutsch. Botan. Gesell., xviii. p. 83, 1900. Translation in J. Roy. Horticult. Soc., xxv. p. 243, 1901.

† Ber. d. deutsch. Bot. Ges., xviii. p. 232, 1900.

‡ Ber. d. deutsch. Bot. Ges., xix. p. 35, 1901; *also* Zeitschr. f. d. landwr. Versuchewesen in Oesterr., iii. p. 465, 1900.

§ Ber. d. deutsch. Bot. Ges., xviii. p. 158, 1900.

¶ Bot. Centralb., lxxxiv. p. 97, 1900.

and *M. glabra*. Thus some of the characters, such as the colour of the flower petals, remained fixed in the hybrids. Also, on the whole, the hybrids had a greater resemblance to the female than to the male parent. In the second generation flowers of new colours, viz., white and red, appeared, in addition to the yellow-white and violet flowers exhibited by the parents. Correns also made a number of crosses between undoubted species (*e. g.*, *Cirsium palustre* + *spinosissimum*, *Achillea macrophylla* + *moschata*, *Carex echinata* + *fœtida*, etc.) and he is doubtful whether one of these hybrids showed a single really dominant character. It was quite obvious that almost all the characters which served to differentiate the parents were present, in greater or less degree, in the hybrids. Correns concludes, therefore, that almost without exception the domination of a character shows itself only in crosses between varieties, whilst the hybrids of true species show the characters of both species, though in diminished degree.

It has been pointed out by Weldon * that Mendel's results are partly vitiated by the fact that he quite neglected the ancestry of the plants with which he started his cross-fertilisations. Weldon also adduces a considerable body of evidence to show that the separation of the seed characters into definite dominant and recessive types by no means invariably holds good. The offspring of cross-bred peas may continue to contain a large percentage of intermediate forms, even as long as 25 generations after the crossing.†

* Biometrica, i. p. 228, 1902.

* For further evidence concerning the Law see Report to Evolution Committee of Royal Society by Miss Saunders and W. Bateson, 1902.

According to Darwin, variability is especially induced if mongrels are repeatedly crossed with either pure parent form, whilst the crossing of different species may lead to much wider variation than the crossing of varieties. The hybrids produced on the first cross are, as a rule, fairly constant in their characters, but if these hybrids be crossed again, or crossed with either pure parent form, then a very considerable variability may result. "He who wishes," says Kölreuter, "to obtain an endless number of varieties from hybrids, should cross and recross them." * Again Darwin † says that cross-bred animals "for breeding are found utterly useless; for though they may themselves be uniform in character, they yield during many generations astonishingly diversified offspring." Indeed it would seem that entirely new characters may be produced by this means. For instance, Kölreuter says that hybrids in the genus *Mirabilis* vary almost infinitely, and he describes new and singular characters in the seeds, anthers, and cotyledons. Professor Lecoq also asserts that many of the hybrids from *Mirabilis jalapa* and *multiflora* might easily be mistaken for distinct species. Again, Herbert ‡ has described certain hybrid Rhododendrons as being unlike all others in foliage, just as if they were a separate species.

According to Focke,§ the hybrid may be related to the parent forms in three different ways: (1) there may

* "Animals and Plants," ii. p. 254.

† *L. c.*, ii. p. 74.

‡ Science Progress, vol. vii. p. 185, 1898.

§ "Die Pflanzen-Mischlingen," Berlin, 1881. Quoted by Weismann, "Germ Plasm," p. 261.

be a strict mean in all parts; (2) the paternal or maternal characters may predominate; (3) the paternal characters may predominate in some parts of the hybrid, and the maternal in others. The first-mentioned condition is by far the most frequent. For instance, Kölreuter states that the hybrid between *Nicotiana rustica* ♀ and *N. paniculata* ♂ (two species of tobacco plant) is exactly intermediate between the parent forms. On the other hand, the hybrid between *N. paniculata* ♀ and *N. vincoeflora* ♂ bears so close a resemblance to the second of these species that the characters of *N. paniculata* can hardly be recognised at all. An instance of the third class is occasionally found in the cross between *N. rustica* ♀ and *N. paniculata* ♂, the blossoms resembling one parent species, and the leaves the other. Again, Milardet * has obtained a series of non-separating crosses by the union of *Fragaria*, *Rubus*, etc. They resembled either the male or the female parent. De Vries † has obtained a similar result with *Œnothera muricata* ♀ × *biennis*, which displayed the paternal character. Crosses between *Œnothera Lamarckiana* ♀ and *O. nanella* ♂ gave progeny which always displayed two types, the maternal and paternal, but these occurred in very varying ratios. Crosses of *O. lata* ♀ and *O. Lamarckiana* ♂ also yielded progeny of both parental types.

Plant hybrids are of considerably more frequent occurrence in nature than animal hybrids, and, by virtue of the fertility which they often possess are of dis-

* Mém. Soc. Sc. Phys. et. Nat. Bordeaux, vol. iv. p. 1, 1894.

† Ber. d. deutsch. Bot. Ges., xviii. p. 435, 1900. *Translation in* J. Roy. Hort. Soc., xxv. p. 249, 1901.

tinct importance as a source of variations. Thus Bennett,* in a paper on Hybridity in Plants, makes the following remark: "There seems, however, scarcely to be room for doubt that in some of our abundant wild genera, such as *Rubus*, *Salix*, and *Hieracium*, hybridity is not uncommon in nature. It has long been known that in some genera, such as *Passiflora*, and in some Orchideæ, the ovules appear to be even more readily fertilised by pollen of a different species. W. Focke now states that this is also the case with the species of *Lilium* belonging to the group *bulbiferum*, and with some species of *Hemerocallis;* and J. H. Wilson affirms the same respecting the Cape genus *Albuca*, also belonging to the Liliaceæ." Again, Rolfe † states that *Narcissus incomparabilis* is known to be wild in France, and that Herbert found that on crossing a daffodil with pollen of *N. poeticus*, the seedlings yielded flowers identical with those of *N. incomparabilis*. Similarly, by crossing the Daffodil with the Jonquil, Herbert succeeded in producing *N. odorus*. Again, Engleheart has proved the hybrid origin of *N. biflorus* by crossing *N. poeticus* with the pollen of *N. tazetta,* he obtaining seedlings identical with wild forms. Also he reconstructed *N. pulchellus Salisb.*, by crossing *N. triandrus* with the Jonquil, the seedlings proving absolutely identical with the wild plant. Further "*Digitalis* supplies some wild hybrids whose origin has been artificially demonstrated. For example, *D. purpurascens*, Roth, has been reconstructed by crossing and recrossing *D. lutea* and *D. purpurea;* and *D. media*,

* Nat. Sci., vol. ii. p. 208.

† J. Roy. Horticult. Soc., xxiv. p. 181, 1900.

Roth, in the same way from *D. purpurea* and *D. ambigua (grandiflora)* . . . *D. ambigua* has also been crossed with *D. purpurea* and with *D. lanata*, in each case yielding hybrids which also occur wild."

Rolfe also records that Wichura succeeded in raising artificially no less than eight hybrid willows identical with those which had long been known in the wild state, and Linton has added at least six others. For instance, *Salix rubra* was obtained by crossing *S. purpurea* with the pollen of *S. viminalis*.

Kerner is of the opinion that species may be produced by hybridisation. In his "Natural History of Plants" he gives instances of these hybrid races. To quote Rolfe, "A hybrid between *Medicago falcata* and *sativa*, known as *M. media*, is widely cultivated as a fodder plant, and is propagated from seed. *Salvia betonicæfolia*, a hybrid from *S. nemorosa* and *nutans*, is as common as its parents in grassland in Central Hungary. *Betula alpestris*, a hybrid between *B. alba* and *nana*, is abundant in the Jura, Scandinavia, and in North Russia, here and there whole copses of it being found. *Nigritella suaveolens*, a hybrid between *N. angustifolia* and *Gymnadenia conopsea*, is abundant in some Swiss localities, hundreds of plants sometimes occurring in a single meadow. Hybrids between the Primrose and Cowslip occur in thousands in upland meadows in the Eastern Alps."

Again, in some localities in the Tyrol the hybrid *Rhododendron intermedium* exists side by side with its parent forms, *R. ferrugineum* and *R. hirsutum*, it sometimes being commoner than they are. Also it seeds freely, and comes true to seed, and so fulfils all the requirements of a species. The same is true of *Salvia sylvestris*, a hybrid from *S. nemorosa* and *pratensis*, which abounds in dry meadows all over the low country south of Vienna, and of *Nuphar intermedium*, a hybrid from *N. luteum* and *pumilum*, which occurs in the Black Forest, Russia, Sweden, and other localities. It appears that a hybrid is sometimes found in company with one parent only, or with one in one locality and both in another; or sometimes even where both are absent.

Kerner estimated that something like a thousand natural hybrids have been found in Europe during the last forty years, but of these hybrids only a fraction survive and multiply.

As regards artificial hybrids, Hurst * has compiled a list of genera from various authorities, and from his own observations, and he finds that 91 distinct genera are recorded in which fertile hybrids are known. In only three, viz., Ribes, Polemonium, and Digitalis, were the hybrids all quite infertile, and in none of them had many experiments been made.

Hurst also remarks that "during the past seven years Mr. Reginald Young has been crossing *inter se* some 30 distinct species and 53 distinct hybrids in the genus Paphiopedilum (Pfitz), and has . . . carefully recorded no less than 849 crosses. Of these, taken together, 80.2 per cent. have proved fertile, *i. e.*, produced good seeds. Of 263 crosses between distinct species, 95 per cent. were fertile. This seems to show that in this genus crosses between distinct species are almost, if not quite, as fertile as crosses between varieties of the same

* J. Roy. Horticult. Soc., xxiv. p. 90, 1900.

species; while in crosses in which a hybrid was concerned in the parentage, out of 586, only 73.5 per cent. proved fertile, showing that crosses with hybrids, though fertile to a high degree, are yet rather less fertile than crosses between species. . . A further analysis of the figures shows that while hybrids crossed with the pollen of pure species give 91.8 per cent. fertile, yet pure species crossed with the pollen of hybrids give but 60 per cent. fertile." That is to say, the decline in the fertility of the hybrids is due in a large measure to the loss of power in the pollen of the hybrids. This decline in power of the male element has been noticed before in other plants by Darwin, Focke, and others.

Rimpau * has made series of experiments on the crossing of some of our common agricultural plants, and, amongst other results, obtained ten artificial and nine natural hybrids in wheat, and two artificial and six natural hybrids in barley. His most striking result of all was to obtain a fertile hybrid between wheat and rye, plants belonging to different genera. Again, Hurst states † that amongst Orchids no less than 150 bigeneric crosses are recorded. Bigeneric hybrids have also been recorded ‡ between *Philesia* and *Lapigeria*, between *Urceolina* and *Eucharis*, between numerous genera of Gesneraceæ, etc. Finally a cross has been described § from *Digitalis ambigua* (Scrophulariaceæ) by pollen of *Sinningia speciosa* (Gesneraceæ); *i. e.*, a *binordinal* hybrid.

* "Kreutzungsproducte landwirth. Cultur-pflanzen," Berlin, 1891.
† *Loc. cit.*
‡ Nature, vol. lxiv. p. 447, 1901.
§ Maund's "Botanic Garden," v. p. 468.

Upon members of the Animal Kingdom very few extensive and systematic crossing experiments have been made. The most complete are those of Standfuss,* on various races and species among the *Lepidoptera*. Standfuss' general conclusion is that on crossing the normal form of a species with a gradually formed local race of the same species, a series of more or less intermediate forms results. For example, on crossing *Callimorpha dominula* ♂ with the variety *persona* ♀, the issue resulting were of a very variable form, more or less intermediate, but somewhat more closely resembling the type than the variety. In the reciprocal cross, the insects, on the whole, also came nearer to *C. dominula* than to the variety, but not so much as before. When species were crossed Standfuss found that the hybrid form lay between the extreme parental forms, but was not strictly intermediate. Arguing from his experiments on crossing various species of *Saturnia*, Standfuss concludes that the adult offspring are more similar to the male parent than to the female, the extent of approximation depending on the relative age of the two species. Crosses of the male hybrids with the parent forms were in some cases proved to be fertile, and hence there is no reason why such forms should not establish themselves under natural conditions. Thus Dr. Dixey, in a very good résumé of Standfuss' researches,† says with reference to these hybrids, " Since the product of this kind of crossing is not found to show a complete reversion to the type of the female parent, it is possible that the existence of vari-

*" Handbuch der paläartischen Gross-Schmetterlinge," Jena, 1896.
† Science Progr., vol. vii. p. 185, 1898.

ous intermediate forms in such genera as *Melitæa*, *Zygæna*, and *Agrotis* may be accounted for in this manner. Cases of simple pairing between distinct species of the two former genera have been observed by the author [Standfuss] in nature."

Upon Echinoids, the author has made numerous crosses and reciprocal crosses.* Eight species were worked with, and of the 56 possible crosses, 41 were attempted. Of these, 22 yielded larvæ of 8 days' growth. In only one cross did any of the larvæ incline towards the paternal type, and the majority of those then obtained were more or less intermediate. In nine other crosses also they were more or less intermediate in character, whilst in the remaining twelve they were of the maternal type. A few of the larvæ exhibited characters which were not present in either parent.

Upon members of the Mammalian and Avian Kingdoms, a very large number of crossing experiments have been made, and frequently with success, but the observations are not sufficiently extensive to admit of generalisations. The most interesting experiments of recent years are those of Professor Ewart, upon zebra hybrids.† By crossing mares of various sizes (11 to 15 hands) with a zebra stallion, nine hybrids were obtained altogether. Also Professor Ewart had in his possession three hybrids out of zebra mares, one having for his sire a donkey, whilst the other two were sired by ponies. The hybrids showed a "curious blending of characters, derived apparently partly from their actual and partly

* Phil. Trans. Roy. Soc., 1898, B. p. 483, *and* Arch. f. Entwickelungsmechanik., Bd. ix. p. 468, 1900.

† "The Penycuik Experiments."

from their remote ancestors. . . Some of the hybrids in make and disposition strongly suggest their zebra sire, others their respective dams; but even the most zebra-like in form are utterly unlike their sire in their markings." In some respects, also, the hybrids were intermediate between their parents.

As to the causes of the different relationships between parental and hybrid characters, we are almost entirely in the dark. Weismann has endeavoured to account for them on his theory of the germ-plasm, but his explanation is purely theoretical and from its nature incapable of experimental verification. The observations of the author on sea-urchin hybrids, and of Professor Ewart on crosses between varieties of rabbits, throw a little light on the subject, for they show that the characters of the hybrids may be considerably influenced by the seasonal condition of the parental sex-cells, and thereby seem to indicate that the comparative degrees of nutrition of the sex-cells, and perhaps also of their constituent parts, may be a very important factor. One should also bear in mind that, as was demonstrated by Mendel in the case of certain plant hybrids, some of the parental characters may remain *latent* in the hybrid offspring, and only reveal their presence in subsequent generations. The existence of latency is also shown by secondary sexual characters. In every female all the secondary male characters, and in every male all the secondary female characters, apparently exist in a latent state, ready to be evolved under certain conditions, such as the removal of the ovaries or testes. The variability of hybrids may therefore be due not only to their having received varying and unequal amounts of the different characters from their parents, to these either partially or entirely refusing to blend, but also to some of the characters received remaining latent, or to characters latent in the parents revealing themselves in the offspring.

Sports. Instances of so-called sports, or suddenly occurring aberrant variations, have been given in the second chapter, but nothing was said of their origin. To what are we to attribute this? Are they to be regarded as normal, only somewhat exaggerated, variations, or are they something essentially different? The more general opinion probably inclines to the latter view, as there are several facts which it is difficult to reconcile with the former. It is said, for instance, that sports, as distinguished from varieties, are much more stable; that they may be transmitted to successive generations with considerable persistence and in undiminished strength. Galton has suggested * that whilst organisms showing ordinary variations are grouped round one "position of organic stability," towards which the offspring in the next generation tend to regress, sports are centred round a different position of stability, and are not merely a strained modification of the original type. They therefore have little tendency to revert to this original type, but are capable of propagating their freshly acquired characters more or less undiminished, and so giving rise to fresh races. Galton considers that the results which he has obtained in his detailed study of human finger-prints † afford strong evidence in support of his view. These patterns, formed by the papil-

* *Vide* "Natural Inheritance," p. 30; *also* "Mind," p. 362, 1894.
† Phil. Trans. 1891, B.

lary ridges on the bulbs of the fingers, are the most persistent of all the external characters that have yet been examined. They are found to fall in three definite and widely different classes. Each of these is a true race in the sense in which that word was defined, transitional forms being rare and the typical forms being frequent. Galton thinks that the continual appearance of these well-marked and very distinct patterns proves the reality of the alleged positions of organic stability.

A clear distinction between sports and varieties seems to show itself also amongst the Lepidoptera. Thus Standfuss * found that when a sport is crossed with its parent form, the issue is sharply divided in both sexes into specimens resembling either the sport or the normal form. There are no true intermediate forms, though occasionally forms are observed in which the characters are unsymmetrically mixed. When the normal form of a species is crossed with a gradually formed local race, however, a series of intermediate forms is obtained. We have seen also that De Vries, in his experiments on plants, claims to have found a wide difference between mere varieties, and true sports such as were obtained from *Œnothera Lamarckiana.*

As already mentioned, sports have been stated to be much more persistent in propagating their aberrant characters than normal varieties, but the evidence in favour of such a generalised statement is quite insufficient. There are certainly a few instances which strongly support it, but there are a good many more which entirely fail to do so. Of the former, the in-

* "Handbuch der paläartischen Gross-Schmetterlinge," Jena, 1896.

stances of the ancon or otter sheep * and japanned or black-shouldered peacocks,† quoted by Darwin, are the most striking. The originator of the ancon breed of sheep was a single ram, born in Massachusetts in 1791. Ancon rams and ewes invariably produced ancon offspring, whilst when crossed with other breeds the offspring resembled either parent, and only very exceptionally yielded intermediate forms. Japanned peacocks, which differ conspicuously from the common peacock in colouring, appear suddenly in flocks of the common kind. Though smaller and weaker birds, they have been known in two instances to increase, and finally extinguish the previously existing breed. They would therefore seem to have been strongly prepotent.

That sports may be no more transmissible than other variations seems to be true in the case of polydactylism, for Dr. Struthers asserts that cases of non-inheritance and of the first appearance of additional digits in unaffected families are much more frequent than cases of inheritance.‡ Again, Galton regards as sports the mental arithmeticians and eminent musicians who are occasionally born into families which in previous generations have shown no signs of such exceptional characters. Though these characters may be transmitted to descendants, yet this is the exception, and not the rule. The subservience of sports to the law of hereditary transmission is well shown by some observations of Standfuss on Lepidoptera. In 1888 a normal female *Aglia tau* was crossed

* "Animals and Plants," i. p. 104.

† *L. c.*, i. p. 305.

‡ *L. c.*, i. p. 458.

with a dark aberrant form or sport of this species, *Aglia lugens*, which had been interbred for two generations. In 1889 some of the *lugens*, both male and female, obtained from this cross, were crossed with normal *tau* specimens. About half the offspring obtained resembled one parent and half the other, intermediate forms being absent. On breeding some of the 1889 ♀ and ♂ *lugens* together, however, their offspring consisted of about 36 per cent. of *tau*, and 64 per cent. of *lugens* forms. In 1890 some of these *lugens* were bred together, and their offspring consisted almost entirely of *lugens*, only 11 per cent. being of the *tau* form. In this latter case, therefore, both parents and all four grandparents were *lugens;* in the 1889 offspring, both parents but only two grandparents, and in the 1888 offspring only one parent and two grandparents.

If sports be of an essentially different nature to normal variations, as the somewhat insufficient evidence available may perhaps be taken to indicate, how is it that they arise? Apparently they occur spontaneously, but doubtless some exciting cause must exist, invisible though it may be. The artificial production of monsters seems to throw some light on the subject, and hence a brief reference to them may be made. These monsters or malformations probably differ from sports only in degree, and not in kind. Hence, if the means adopted for their artificial production are such as may occur under natural conditions, it seems possible, and even probable, that sports themselves may owe their origin to similar agencies. For instance, Dareste, as long ago as 1877, described numerous experiments on the effects of placing fowls' eggs vertically instead of horizontally during development, of keeping them slightly above or below the normal temperature of incubation, of heating different parts of the egg unequally, and of modifying the conditions of respiration by varnishing part of the shell.* Various considerable malformations were produced, but these were more or less the same, whatever the conditions that produced them. Professor Windle,† who has extended these investigations and determined the effects of various other agencies, as electricity and magnetism, on development, came to a similar conclusion. He considered that these disturbing agents act, in the majority of cases, on that part of the developing organisation which is concerned with the formation of the vascular system of the embryo, and so indirectly produce the malformations observed.

The suggested connection between considerable malformations and sports has not as yet been borne out by Dareste's researches, although observations have been made with the object of finding it. In these observations the conditions found to produce considerable malformations were reduced in strength, in the hope of thereby obtaining only slight anomalies, compatible with continued existence and the procreation of offspring. Unfortunately the domestic fowl, which was invariably made use of, is unsuitable for such observations. The type is so diversified that the experimenter who obtains some variety can never be certain whether

* "Recherches sur la Production artificielle des Monstrosités," Paris, 1877; second edition, 1891.

† Proc. Birmingham Phil. Soc., vii. p. 220, 1890.

it should be attributed to the conditions of experiment, or to some physiological cause arising in the egg itself. To test this question with some chance of success, the eggs of some species which varies but little ought to be employed; *e. g.*, some wild species. But in this case it would be very difficult to obtain sufficient material.

In the case of certain Lepidoptera, however, the artificial production of sports has been successfully accomplished by Standfuss.* By keeping the pupæ of *V. cardui* (Painted Lady) at a high temperature for a short period, he succeeded in producing a small number of specimens of the aberrant form *elymi*, a form which is occasionally found under natural conditions. Again a low temperature, acting on pupæ of *V. io* (Peacock), produced a variety ab. *fischeri*, which exhibits a reduction in the number of the blue scales on both fore and hind wings. In these and other characters there seemed to be an approach to the type of *V. urticæ*. These and other observations seem to justify Standfuss' conclusion that many of the aberrations occurring in nature may likewise have arisen through the influence of abnormal temperature conditions.

Telegony. The term *telegony*, or so called *infection of the germ*, is applied to certain cases apparently showing the influence of a previous fertilisation on the structure of the subsequent offspring. The test case, always quoted in support of the existence of this phenomenon, is that of Lord Morton's mare. This animal bore a hybrid to a quagga, and subsequently produced two colts by a Black Arabian horse. These colts, both in the hair of their manes, their partial dun colour, and

* The Entomologist, vol. xxviii. p. 145, 1895.

striping on the legs, strongly resembled the quagga. Other cases have been quoted in support of this phenomenon, but it is unnecessary to mention them here, for none of them are absolutely convincing. The reader who wishes for details of these cases should consult a useful paper by Finn.* In his Penycuik experiments, Professor Ewart has made a number of attempts to obtain evidence of the phenomenon, but so far with entirely negative results. Sir Everett Millais made a considerably larger series of experiments, on a variety of animals, but was equally unsuccessful. Many German breeders also believe telegony as yet unproven. Finally, Professor Pearson † has shown that exact statistical examination of appropriate data gives no support whatever to the hypothesis. Pearson's method of testing the question was to determine whether younger children are more closely correlated to their parents in respect of some character such as stature, than older children. Supposing the male parent were able to exert any influence on the maternal tissues, and so indirectly on the offspring, then clearly this influence would be greater for the younger children than for the older children. As Pearson recognises, it is possible that telegony, if it occurs at all, is due to the abnormal preservation of the male sex cells of an earlier union, and in such a case his method would afford no evidence one way or the other.

Probably, therefore, no such thing as telegony exists. In any case it is so exceedingly rare that, as a possible source of variations, it may be neglected.

* Nat. Sci., vol. iii. p. 436.

† Proc. Roy. Soc., lx. p. 273.

Parthenogenesis. We saw in the last chapter that Weismann regarded sexual reproduction as a potent factor in the production of variations, in that it afforded inexhaustible supplies of fresh combinations of the individual variations already represented in the mingling germ-plasms. We should accordingly conclude that when such sexual union is wanting, as in parthenogenetically produced animals, the amount of variation will be smaller, and that parent and offspring will more closely resemble each other. The evidence upon this point is exceedingly slight, but what there is perhaps tends rather to support this deduction. Thus Weismann made a series of observations, extending over eight years, upon a small ostracod, *Cypris reptans.* This organism exists as two well-marked varieties, one being coloured yellow, with five small green spots on each side of the shell, and the other seemingly dark green, owing to the great enlargement of these spots.* Both varieties are produced parthenogenetically in the neighbourhood of Freiburg, males never being found. Females of each variety were isolated, fed well, and allowed to multiply for many generations. It was found that "the descendants of the same mother resembled one another as well as the parent with which the experiment began, even as regards minute details of the markings. The differences were mostly as small as those which may be observed in identical human twins." Even after many generations no modification showed itself, so that colonies were obtained which could not be distinguished from their ancestors 40 generations back. In three different instances, however,

* "Germ-Plasm," p. 344.

some of the dark green variety appeared in broods of the typical yellow variety, and in one instance some of the yellow variety in broods of the dark green. These sudden transformations could not have been due to external circumstances, as the two forms appeared in the same aquaria, under precisely the same conditions. Weismann attributes them to reversion.

Evidence telling in the opposite direction to Weismann's has recently been obtained by Warren,* and as it is based on exact statistical measurements, one is inclined at first sight to attach greater weight to it. The observations consisted in measurements of the total length of body to base of spine, and of the length of the protopodite of the second antenna of the right side, in 23 female *Daphnia magna*, and their 96 parthenogenetically produced offspring. As these animals continue to grow throughout life, the second dimension was expressed in terms of the first, before calculating its variability. Its error of mean square, or standard deviation, was found to be 2.22 in the mothers, and 2.95 in the offspring. That is to say, the offspring were distinctly more variable than the mothers, and even the offspring of a single mother were found to be on an average more variable than all the mothers put together. As the mothers had in a way been selected, only those which produced offspring being chosen, the daughters would be expected to be somewhat more variable, but in any case the variability was considerable. Again, it was found that the coefficient of correlation between mother and offspring was .446, whilst the coefficient of regression of offspring on mothers was .619.

* Proc. Roy. Soc., lxv. p. 154, 1899.

Now it has already been found, in the case of stature in man, that the correlation between mid-parent (*i. e.*, mean between male and transmuted female) and offspring is .424, and the regression of offspring on mid-parents .6;* hence Warren's values seem to show that the parthenogenetic mother acts as a mid-parent to her offspring, and not as a single parent, and also that these offspring exhibit regression towards the mean race type, just as sexually produced individuals do. As Warren himself points out, however, his evidence is not conclusive. Thus the number of individuals measured was comparatively small, and also it would seem that Daphnia is a very unreliable organism to work with. It is so exceedingly sensitive to its environmental conditions †—very considerable variations being produced by comparatively slight changes—that these data derived from it can only be accepted with considerable reserve.

A further series of observations was made by Warren ‡ upon Aphides (*Hyalopterus trirhodus*). Sixty parents and their 368 children were measured, and also 30 grandparents and their 291 grandchildren. Warren found that the coefficients of parental and grand-

* It has been stated in the previous chapter that the coefficients of correlation and of regression between single parent and offspring are practically the same thing, and are equal to .3. The coefficient of correlation between mid-parent and offspring is, however, $\sqrt{2} \times .3 = .424$, because the mid-parent, being the mean of two parents, is less variable than the single parents (in the proportion of 1 to $\frac{1}{\sqrt{2}}$). The coefficient of regression of offspring on mid-parents, is, however, twice that of offspring on a single parent, *i. e.*, is .6.

† *Vide* Q. J. Microsc. Sci., vol. xliii. p. 199, 1900.

‡ Biometrika, i. p. 129, 1902.

parental correlation showed no marked difference from those obtained in sexual reproduction, just as in the case of his daphnia observations. If anything, there was a decrease in the correlation, on passing from sexual to parthenogenetic forms, rather than the increase we should expect. However, the variability of the individuals of a brood was found to be only about 60 per cent. of the racial variability; *i.e.*, distinctly less than in sexual reproduction. Also the mean coefficient of fraternal correlation for aphis and daphnia was .66, or considerably higher than the mean value of .45 obtained by Pearson * for fraternal correlation among sexual forms. Warren's general conclusion may be summed up in the words: "The question as to whether we have a real difference between parthenogenetic and sexual offspring can only be decided by further investigation both on aphis and other forms." In the light of Weismann's observations, which were carried on for such a number of generations, we seem entitled to conclude that probably a real difference will be found to exist between them.

Arguing partly from Warren's observations, and partly from others of his own, Professor Pearson † has come to the conclusion that, "whatever be the function of sex in evolution, it is not the production of greater variability." Thus he says that the individual contains in itself a variability which is 80 to 90 per cent. of the variability of the race, and which it can exhibit quite independently of sexual union, *e. g.*, as in this case of parthenogenesis. As instances of individual varia-

* Phil. Trans. 1901. A. p. 285.

† "Grammar of Science," p. 474.

bility, he refers to the stigmatic bands on the seed capsules of Shirley poppies. These vary in number from about 7 to 18, the most commonly occurring number being 12. The variability of a large number of individuals (as expressed by the error of mean square), which were taken as a good sample of the whole race, was found to be 1.885. The average variability of the bands in the capsules obtained from each of 300 different plants, or the individual variability, was, however, only 15 per cent. less. Again, with reference to the number of leaflets on the compound leaf of the ash, the individual variability was only 8 per cent. less than the racial variability.

In a recent paper,* Professor Pearson and his co-workers have determined the relationship between racial and individual variability in a number of other plant species. Enumerations were made of the veins in the leaf of the Spanish Chestnut and the Beech, of the prickles on Holly leaves, the sori on the fronds of Hartstongue ferns, the seeds in the pods of Broom plants, etc., and measurements of the length and breadth of ivy leaves and of the gills of mushrooms. On an average, the individual variability was found to be about 87 per cent. of the racial, it varying in the different series of observations between 77 and 98 per cent. Now, even admitting that in these instances the individual variability is only slightly smaller than the racial, it does not appear to me that Professor Pearson is entitled to his contention, for all these highly variable individuals were, of course, produced as the result of sexual union in their immediate or remote progenitors.

* Phil. Trans, 1901, A. p. 285.

Such union may have been the starting point of considerably increased variation, which was never lost, even through innumerable subsequent asexual generations. Thus Professor Pearson has shown that if the ancestors of individuals be selected so as to be absolutely similar in character for an indefinite number of generations back, such individuals will still have a variability of upwards of 89 per cent. of that of the original race. Though produced sexually, these individuals are in reality comparable to asexually reproduced forms, as by hypothesis no new characters were introduced by any of their ancestors.

Whether the difference between racial and individual variability is as small as Pearson maintains, or not, depends solely on what is meant by the word "race." If "species," in the generally accepted sense, is meant, then the view is certainly incorrect. If, however, a group of individuals is implied, all of which have been exposed during several generations to practically identical conditions of environment, then the view must be admitted. It is of little practical value, however, as may be realised by quoting certain data which Pearson has himself brought forward in another connection.* Thus the variability in the number of stigmatic bands in a sample of the wild poppy, *Papaver Rhœas*, collected in a corn field at the foot of the Chiltern Hills, was found to be 1.473, that in two individual poppy plants being, on an average, 1.166, or 20.9 per cent. less. Another sample was collected in some fields at the top of the Chilterns, and in this case the variability was 1.770, or 20.2 per cent. greater than in the other sample. Moreover, the

* "Grammar of Science." p. 387.

mean number of bands was also greater, it being 10.04 as against 9.84. In a third sample collected from still another locality, the variability was 1.455, but the mean number of bands was only 8.77. Supposing, therefore, equal numbers of specimens had been collected from all three localities and combined, the variability would have been about double the average variability of the individual groups of plants. Supposing samples had been collected from numerous and more widely separated localities, so as to get a representative sample of the whole species, then doubtless the variability would have been much greater still. Individual variability may therefore be only slightly smaller than local racial variability, but it is very much smaller than specific variability.

What is true for plants is true also for animals. Supposing that in the case of the middle classes of English society, the average variability of the stature of all the offspring is only about 10 per cent. more than that of the offspring of individual parents, then it is clear that if we were to include also representatives of the lower and of the upper classes in our sample, the average variability would be somewhat greater, perhaps 12 per cent. If we were to include representatives in due proportions from all the continental nations, then the variability might be 25 per cent. or more in excess, and if from all the nations of the world, with African pygmies on the one hand, and Patagonian giants on the other, then it might be 50 per cent. greater, or even more.

Asexual Reproduction in Plants. In plants asexual reproduction is very much more common than in animals, and though, until the above-mentioned memoir was published, there was practically no statistical evidence as to the range of variation then experienced, as compared with that found in sexually reproduced forms, there was available the common knowledge of every horticulturist and nurseryman, were he scientifically trained or otherwise. Thus it had been thoroughly well established that asexually produced forms, *i. e.*, grafts, cuttings, offsets, and tubers, are characterised by a very much greater constancy than sexually produced forms, *e. g.*, seedlings. For concrete instances I cannot do better than quote from those given by Mr. Sedgwick in his recent Presidential Address before the British Association.* For example, in the asexual propagation of the potato by tubers, the plants, be they, for instance, of the Magnum Bonum variety, give rise to plants exactly resembling their parent in foliage, flowers, and tubers; if they be of the Snowdrop variety, the foliage, flowers, habit, and tubers are also similar, and are totally different from those of the Magnum Bonum. "In this way a favourable variety of potato can be reproduced to almost any extent with all its peculiarities of earliness or lateness, pastiness or mealiness, power of resisting disease and so forth. By asexual reproduction the exact facsimile of the parent may always be obtained, provided the conditions remain the same." Supposing, on the other hand, one tries to raise Magnum Bonum plants from seed, in all probability not one of the seedlings will exactly reproduce the parents; they will all be different, both in properties of keeping, resisting disease, and so forth. "Again, take the apple:

* Nature, vol. xl. p. 502, 1899.

if you sow the seed of the Blenheim Orange and raise young apple trees, you will not get a Blenheim Orange. All your plants will be different, and probably not one will give you apples with the peculiar excellence of the parent. If you want to propagate your Blenheim Orange and increase the number of your trees, you must proceed by grafting or by striking cuttings."

In the face of such evidence as this, it seems impossible for Professor Pearson to maintain his belief that the function of sex in evolution "is not the production of greater variability." At the same time, his results above quoted show the incorrectness of the view sometimes held, that variability is quite insignificant in asexually, as compared with that in sexually, reproduced forms. Statistically measured, it is only 10 to, say, 50 per cent. less, though when this amount is translated into differences of foliage and flowers, or of quality of fruit, it seems at first sight much more considerable.

Bud-Variation. Considerable variations may arise asexually in cases of so-called bud-variation. This term was used by Darwin to designate the sudden changes in structure and appearance which occasionally occur in the flower-buds or leaf-buds of full-grown plants. Such changes are known to gardeners as sports, but, as we have already seen, this term is now used to include all suddenly arising discontinuous variations.

One of the best known and most striking instances of bud-variation is that of the nectarine, which occasionally appears on full-grown peach trees which have previously borne peaches alone. This is the more remarkable as most varieties of both the peach and the nectarine reproduce themselves truly by seed. Again, nectarine stones occasionally yield peach trees, and a single instance is recorded of a full-grown nectarine tree bearing perfect peaches.* Numerous other instances of bud-variation have been observed in the plum, cherry, vine, gooseberry, currant, and other fruits, but it is unnecessary to refer to these here. In flowering plants, also, many cases have been recorded of a whole plant, or a single branch or bud, suddenly producing flowers different from the proper type in colour, form, size, or other character. For instance, a Chrysanthemum, raised from seed, produced by bud-variation six distinct varieties, five differing in colour, and one in foliage.† The common double moss-rose probably took its origin from the Provence rose by bud-variation. The leaves and shoots may be modified by bud-variation as well as the flowers, and several varieties of trees have probably originated in this manner.

As to the cause of bud-variation, we are in the majority of cases entirely ignorant. Darwin attributes many of the cases to reversion to characters previously present, but which have been lost for a longer or shorter time. Other cases he attributes to the plants being of crossed parentage, and to the buds reverting to one of the two parent forms. There are still many cases left, however, in which what appear to be absolutely new characters present themselves. These can only be at-

* "Animals and Plants," i. p. 362.
† "Animals and Plants," i. p. 440.

tributed to so-called spontaneous variability. Though in individual instances it may be difficult or impossible to assign any reason for the sudden appearance of such "spontaneous variability," yet various observations incline one to believe that it is probably, after all, only a special instance of variations due to changed conditions of life. Thus it is noticeable that all plants which have yielded bud-variations have likewise varied greatly by seed. They seem, in fact, to possess an inherent variability. Again, almost all plants showing bud-variation have been highly cultivated for long periods, in many soils and under different climates. On the other hand, plants living under their natural conditions are very rarely subject to bud-variation. In some instances, as when all the fruit on a purple plum tree suddenly becomes yellow, or all the fruit on a double-flowered almond suddenly becomes peach-like, we seem to perceive a direct result of changed conditions of life; but more often than not we are compelled to conclude that the connection is only an indirect one.

Darwin points out * that it is "a singular and inexplicable fact that when plants vary by buds, the variations, though they occur with comparative rarity, are often, or even generally, strongly pronounced." In plants raised from seed, however, the variations are almost infinitely numerous, but their differences are generally slight. Bud-variations clearly seem, therefore, to be true discontinuous variations, and not merely exaggerated normal variations. As to the ultimate cause of their production, we are as completely in the

* *L. c.*, i. p. 443.

dark as we were for the analogous phenomena observed in sexually produced forms. We can only conclude, as we did then, that the process of development at some point takes on a new and abnormal departure, the direct or indirect result of changes of environmental conditions.

CHAPTER VI.

CERTAIN LAWS OF VARIATION.

Effect of environment on growth diminishes rapidly from time of impregnation onwards—Reaction of an organism to environment dependent on nature of organism—Rapidly diminishing rate of growth in man and in the guinea-pig with progress in development—Variability also diminishes with growth—Effect on growth once produced, probably never eradicated—Increased variability of sparrow and of periwinkle in America—Relation between variability and want of adaptation to environment—Variability of migratory and non-migratory birds—Does domestication increase variability?

Before entering on the discussion of the causes of so-called somatogenic variations, *i. e.*, of acquired characters, it will be well to examine at some little length certain more or less general laws and conditions which control their acquisition and retention. This is the more necessary, as the matter has received such very little attention hitherto. It seems to have been more or less tacitly assumed that external conditions act equally powerfully at all periods in the growth of a developing organism, whilst the persistence or otherwise of any effect, once produced, has scarcely been debated at all.

In what way, then, does a developing organism react in its growth to the conditions of its surroundings? It would probably be concluded that any given change of environmental condition would produce more effect in the earlier stages of development than in the later, but what is the numerical expression of this difference? Such an expression can be obtained in two ways: directly, as the result of experiment; and indirectly, from certain considerations as to the rate of growth and persistence of variations.

To determine the effect of environment on growth, almost any organism can be made use of, but it would obviously be exceedingly troublesome and laborious to work with the higher organisms, such as mammals. In them the earliest stages of embryonic development would be especially difficult to reach. With many of the lower organisms, however, all such difficulties are avoided, and an inexhaustible supply of material can readily be obtained. For these reasons the author attempted to investigate the question at issue by observations on the larvæ of sea-urchins. The method of experiment has already been indicated in Chapter IV., so that it is unnecessary to refer to it again.

The object in view was to determine the permanent effect of some abnormal environmental condition, acting at various periods of development, on the size of the larvæ. It was found that their growth practically ceased after 6 or 8 days, and hence any effect then found to be present was fixed and ineradicable, so far as the larval stage of the organism was concerned. The most convenient environmental condition to work with proved to be temperature, for it is easily controlled, and the effect produced may be considerable. Thus larvæ kept at 10° C. during growth are some 25 per cent. smaller than those kept at 20° C.*

*Phil. Trans. 1898, B. p. 481.

To determine the effect of temperature acting at the time of impregnation, portions of the ova and spermatozoa were shaken from the ovaries and testes in small beakers of sea-water. After bringing these to the required abnormal temperature, their contents were mixed, and the mixed solution kept at the same temperature for, in most cases, an hour. It was then poured into a jar holding 2 to 4 litres of sea-water at the normal temperature. The ova, now fertilised, were allowed to develop under as constant conditions as possible for 8 days, and the larvæ were then killed and measured in groups of 50, as already mentioned. Other ova, kept at the time of impregnation at a normal instead of an abnormal temperature, were allowed to develop under otherwise exactly similar conditions, and so afforded "control" or "normal" larvæ, against which the effect produced in the other larvæ by exposure to the abnormal temperature could be determined. In the accompanying table the results obtained in the various observations are collected: *

TIME OF EXPOSURE TO ABNORMAL TEMPERATURE.	ABNORMAL TEMP.	NORMAL TEMP.	PERCENTAGE DIMINUTION PRODUCED IN SIZE OF LARVÆ.	AVERAGE EFFECT.
1 hour	about 8°	about 19°	8.5, 1.8, 8.3, 0.0, 1.2, 2.5, 8.7, 2.5, 3.6, 4.2.	4.1%
"	" 25.5°	"	3.8, 13.8. 4.7. + 1.0, 5.1, 9.4, 6.0.	5.9
1 or 8 min.	" 8°	"	.3, 6.9, 3.6, 2.4.	3.3
" "	" 25.5°	"	10.6, 1.9, 2.7.	5.1
10 seconds	8° or 25.5°	"	2.4, 2.7, + .2, 1.9.	1.7

Here we see that exposure for an hour to a temperature of about 8° C. at the time of impregnation, instead

* *Vide* Phil. Trans. 1895, B. p. 582, and Proc. Roy. Soc., vol. xlvii. p. 85, 1900.

of one of 19° C., produced, in ten observations, an average diminution of 4.1 per cent. in the size of the larvæ. Temperatures a few degrees above the normal acted even more unfavourably, one of 25.5° producing, in seven observations, an average diminution of 5.9 per cent. It follows, therefore, that at the time of their impregnation, the ova are most extraordinarily sensitive to the temperature of their surroundings, be it abnormally high or abnormally low. Further observations showed that they were also very sensitive to another condition, viz., salinity of the water, though not to the same extent as to temperature. It seems very probable, therefore, that at this period they are very sensitive to all conditions of environment, whatever their nature.

The results contained in the lower half of the table are even more remarkable than those in the upper. Thus, if the ova were kept at about 8° or at 25.5° for only one to three minutes after the mingling of the solutions containing the ova and spermatozoa, and after this short period were poured into jars of water at normal temperature, an average diminution in size of respectively 3.3 and 5.1 per cent. was effected! The observations made are not very numerous or regular, and hence not much importance can be attached to the actual figures, but one is justified in concluding that the effect produced is not so very much smaller than when the period of exposure to the abnormal temperature was an hour. Now during a minute's time little more than the impregnation of the ovum would be accomplished. The processes connected with the fusion of the male and female pronuclei and the commencement of segmentation can scarcely have begun; hence it must be the temperature of the ovum and the spermatozoön during the act of impregnation which is of such considerable influence on the subsequently developing larvæ, the effect perhaps depending on the vibratile energy with which the spermatozoön strikes and enters the ovum. In the last line of the above table are recorded a few experiments in which the time of exposure to the abnormal temperature was reduced to ten seconds. The average effect then produced was only 1.7 per cent., or less than half as much as in the other observations. Presumably, therefore, the time was insufficient for all the ova to become impregnated at the abnormal temperature.

It may perhaps be thought that this extraordinary sensitiveness of the ova to temperature at the time of impregnation is true only for the particular species of sea-urchin worked with, viz., *Strongylocentrotus lividus*, and that one is in no way justified in regarding it as a general phenomenon. How far this criticism is true or not, future experiment alone can show, but the sensitiveness was at least demonstrated in one other case, viz., the ova of the sea-urchin *Sphærechinus granularis*. Exposure of the ova for one hour to a temperature of about 11° produced in four experiments respectively 7.7, 1.1, 4.6 and 2.6 per cent. diminution in the size of the eight-days' larvæ. Exposure to a temperature of 27° was much less effectual, it producing in two experiments respectively 2.5 and .9 per cent. diminution. In two other experiments, exposure to this temperature for a minute instead of an hour produced respectively 4.3 and .9 per cent. diminution. In explanation of these slighter effects it should be mentioned that this organism throughout its growth was found to be distinctly less reactive to its environment than was *Strongylocentrotus*.

In order to determine the effect of exposure to abnormal temperatures during the later stages of development, all the *Strongylocentrotus* ova were kept for the first hour during impregnation at the same temperature, and were then divided up into two portions. One portion was poured into a jar of water at a normal temperature, and the larvæ obtained therefrom used as control larvæ, whilst the other was poured into water previously brought to an abnormal temperature. After a few hours the contents of this jar were vigorously stirred, and some of them poured off into a smaller jar. This was then transferred to the tank of running water containing the control jar, and kept there during the remainder of larval develpoment. A few hours later another portion was withdrawn from the abnormal temperature jar, and similarly treated. The kind of result thereby obtained may be gathered from the following table:

	SIZE.	PER CENT. DIMINUTION IN SIZE PER HOUR.	DURING
Normal larvæ (22.5°),	100.0		
1— 6 hrs. at 12°,	93.61	1.28	1— 6 hrs.
1—10 hrs. "	92.37	.31	6—10 hrs.
1—21 hrs. "	90.09	.21	10—21 hrs.

Here the size of the "normal" larvæ, grown throughout at a temperature of 22.5°, is taken as 100. Larvæ obtained from ova kept at a temperature of about 12° from the end of the first hour after impregnation to the end of the sixth hour were found to be

6.4 per cent. smaller than these, or were, on an average, diminished in size 1.28 per cent. for each hour's exposure. An exposure of the ova for a further four hours produced only 1.24 per cent. more diminution in the size of the larvæ, or .31 per cent. per hour, whilst still eleven hours more produced an additional 2.28 per cent. diminution, or .21 per cent. per hour. Taking means of the various hours of exposure, one can therefore say that the effect produced during hour 3½ was 1.28 per cent., or about four times greater than that produced during hour 8, and six times greater than that in hour 15½. That is to say, the effect produced very rapidly diminished with progress in development.

NORMAL LARVÆ AT ABOUT 20°: ABNORMAL AT 8°.			NORMAL LARVÆ AT 13°: ABNORMAL AT 22°.		
Time of exposure in hours.	Mean time in hours.	Per cent. diminution in size per hour.	Time of exposure in hours.	Mean time in hours.	Per cent. increase in size per hour.
0– 1	.5	4.14	1–8	4.5	1.08
1– 6	3.5	1.28	1–11	6	.0
1– 9	5	1.17	8–19	13.5	.37
6–10	8	.31	1–28	14.5	.40
10–21	15.5	.21	19–43	31	.0
			1–84	42.5	.13
			28–71	49.5	.125
			19–192	105.5	.0
			84–192	138	.022

Other observations were made at a colder season of the year, and in these the normal larvæ were kept at about 13°, and the "abnormal" at about 22°. An increase in the size of the larvæ was of course produced thereby, and here again the effect was very much greater for exposure during the earlier hours of development than during the later. In the foregoing table are collected all the observations made by both methods, the mean effect above mentioned as being produced by an hour's exposure to a temperature of 8° at the time of impregnation being also included. We accordingly see that the unfavourable influence of the cold steadily diminishes from the time of impregnation up to the 15th hour; whilst in the other series of observations the favourable influence of the warmth also diminishes rapidly, though not so regularly. In one or two cases, for some unknown reason, apparently no effect was produced; but allowing for these by taking means of the values obtained at more or less similar hours, we still find a rapid and fairly regular decrease. Thus, at about the 120th hour the average effect produced was only about a fiftieth of that at the fifth hour.

In still a third series of observations, some of the ova were kept for various periods at about 26° C. During the first few hours of development this temperature had an extremely unfavourable action, producing a diminution in size varying from 20.76 to 7.36 per cent. Still further exposure to it, however, produced a favourable effect on growth, though this was never sufficient entirely to counteract the previous adverse influence. The results obtained in two of the experiments made are given in the table below. Here we see that from the end of the first to the end of the fourth hour average diminutions of 3.89 and 6.45 per cent. per hour were produced. During the next four hours this adverse effect dwindled down to nothing, or almost nothing, and in the next four hours a distinctly positive effect set in. This probably continued for the remainder of the period of growth. If we neglect the transition periods in these two series, we see that the effect of the abnormal temperature, be it negative or positive, rapidly diminishes with progress in development. In the 17th hour it is not a fifteenth and in the 83rd hour not a four hundredth as great as in hour 2½. In two other series of observations a similar kind of result was obtained; so taking together all the series of observations made, by all the methods, one is therefore justified in concluding that the effect of temperature on the growth of the developing organism diminishes rapidly and regularly from the time of impregnation onwards.

	SIZE OF LARVÆ.	PER CENT. VAR. PER HOUR.	DURING	SIZE OF LARVÆ.	PER CENT. VAR. PER HOUR.
Normal larvæ (at 23.5° and 24.2°)	100.00			100.00	
1–4 hours at 26°	88.33	−3.89	1–4 hrs.	80.64	−6.45
1–7½ or 8 hrs. "	88.90	+ .14	4–7¾ hrs.	79.24	− .40
1–11 or 12 hrs. "	94.56	+1.42	7¾–11½ hrs.	80.21	+ .28
1–22 hrs. "	99.31	+ .47	11½–22 hrs.	84.25	+ .20
1–144 hrs. "	98.43	− .007	22–144 hrs.	87.27	+ .025

These experiments in which the same condition of environment at first produced an adverse effect, and then a favourable one, illustrate a very important principle; one which was recognised long ago by Darwin, and also by Weismann. They show that the reaction of an organism to its environment depends on the nature of that organism. Thus with respect to the direct action of conditions of life, "We must bear in mind," says Darwin, "that there are two factors; namely, the nature of the organism and the nature of the conditions. The former seems to be much the more important; for nearly similar variations sometimes arise under, as far as we can judge, dissimilar conditions; and on the other hand, dissimilar variations arise under conditions which appear to be nearly uniform." * The reasons of these differences of reaction are, as a rule, quite unknown to us, but in the present instance a satisfactory explanation was arrived at. Thus, by heating up portions of the ova to various temperatures at various periods of development, it was found that the temperature at which they were killed was by no means constant. It increased regularly with progress in development, so that it was about 12° C. higher for full-grown plutei than for ova at the time of impregnation.

STAGE OF DEVELOPMENT.	TIME AFTER IMPREGNATION.	DEATH TEMPERATURE.
Strongylocentrotus ova,		28.5°
Semi-blastulæ,	4 hours	33.5°
Blastulæ and semi-gastrulæ,	12 hours	36.5°
Plutei and semi-plutei,	28 hours	39.5°
Plutei,	6 days	40.3°

Supposing, therefore, a temperature of 33.5° is fatal to four hours' blastulæ, then all temperatures a few degrees lower than this must be unfavourable to their growth. Lower temperatures still we know to be highly favourable. Now we saw that a temperature of 26°—*i. e.*, one 3° or 4° below the death temperature—was very unfavourable to the growth of ova shortly after impregnation; so to produce an equally unfavourable effect on four hours' blastulæ, we should probably need a temperature of 30° or so, and on 12 hours' blastulæ, one of 33°. Or conversely, the temperature of 26,°

* "Origin of Species," Ed. 6, p. 6.

which proved itself so favourable to the growth of 12 hours' blastulæ, would correspond to a temperature of 23° acting on 4 hours' blastulæ, and one of about 19° acting on ova shortly after impregnation.

To return to our conclusion as to the diminishing effect of temperature with development, it seems reasonable to imagine that what is true for one environmental condition is probably true for others also. Observations were in fact made to test the effects of a change of salinity, and they also proved the ova to be very much more sensitive in the earlier stages of their growth than in the later.* Here again, however, a double effect was produced, adverse for the first few hours of exposure, and favourable for the later ones.

Due reflection will, I believe, incline one to infer that what is true for Echinoid larvæ is true for most multicellular organisms. In fact, it would seem to be a law of general application that *the permanent effect of environment on the growth of a developing organism diminishes rapidly and regularly from the time of impregnation onwards.*

It is curious that this principle, enunciated by the author in 1900, should have been laid down by De Vries,† only a few months later, as the result of his observations on plants. Thus, judging from the effects of nutrition (manuring, planting out, good light, and watering), he concluded that: (1) The younger a plant is, so much the greater is the influence of external conditions on its variability. (2) The nutrition of the

* *Vide* Proc. Roy. Soc., vol. lxvii. p. 97, 1900.

† "Die Mutationstheorie," p. 373.

seed, when developing on the maternal plant, has—at least very often—a greater influence on the variability than nutrition during its germination and later growth.

De Vries does not quote any concrete instance in support of the first of these conclusions, but in proof of the second, he adduces some observations on *Œnothera Lamarckiana.** On sowing some seed of this plant in well-manured ground, he found that the seed capsules of the plants obtained were slightly smaller than those from seeds grown in unmanured ground (the mean lengths being respectively 25.2 and 27.2 mm.). The seeds of three of the manured plants, when sown next year, yielded plants with seed capsules, on an average, 29.9 to 33.4 mm. in length. This considerable increase seems to have been almost entirely due to the manure treatment received by the plants from which the seeds had been derived, and in comparison with it, artificial selection was not nearly so effective. Thus the (well-manured) plants arising from seeds taken from long seed capsules (32.6 to 43.0 mm. in length) had themselves capsules of only 31.6 to 33.4 mm., whilst those from the seeds of short seed capsules (15.6 to 23.4 mm.), had capsules of 24.2 to 29.9 mm.

Doubtless the numerical measure of the effect of environment on growth varies enormously in different organisms, and the extent of its variation can be determined only by experiment. Fortunately, however, numerous data which have been collected as to the rate of growth in certain animals, more especially in man and the guinea-pig, can be readily applied to the partial

* *Loc. cit.*, p. 383.

elucidation of the law in question, though they are insufficient to afford an absolute criterion of its extent. Thus it is obvious that the reaction of a growing organism to its environment must depend on its rate of growth at the time. Supposing that in the earliest stages of development it doubles its weight within a day, whilst in the later stages only within a month, then an unfavourable environmental condition, acting for a given time during the early period, may conceivably produce thirty times more retardation of growth than when acting during the later period. To what degree retardations of growth produced at various stages of development persist to the adult stage is another question, and reference will be made to it later on.

In the case of man, the data relative to the embryonic rate of growth are not very reliable, but they are quite sufficiently so for our purpose. A considerable number of them have been collected by Preyer.* His has furnished data as to the length of embryos during the first few weeks of development, and Hennig has given a table of measurements for the 1st to the 10th month, these being averages derived from a hundred observations. As Minot points out,† however, these data are inexact, for in the early stages only the head and trunk were measured, and in the later stages head, trunk, and legs. In human embryos weight is a better criterion of development than length, but as in the observations of His only the length was determined, it is necessary, for purposes of comparison, to keep to this standard throughout. In order to obtain readily comparable

* "Specielle Physiologie des Embryo," p. 500, Leipsig, 1885.
† "Human Embryology," p. 336, New York, 1892.

values, the times required by the embryo at the various stages of development in order to double its length have been calculated:

PERIOD OF DEVELOPMENT.	LENGTH DOUBLED DURING	PERIOD OF DEVELOPMENT.	LENGTH DOUBLED DURING
$2\frac{1}{4}$—$2\frac{3}{4}$ weeks	6.6 days	4—5 months	36.7 days
$2\frac{3}{4}$—$3\frac{1}{2}$ "	9.5 "	5—6 "	78.2 "
$3\frac{1}{2}$—4 "	7.8 "	6—7 "	146.3 "
1—2 months	11.6 "	7—8 "	202.5 "
2—3 "	26.2 "	8—9 "	306.1 "
3—4 "	29.6 "	9—10 "	518.5 "

From this table we see that the rate of growth of the human embryo diminishes steadily, with one slight exception, from the earliest to the latest stages. So enormous is the diminution in the growth that in the last month of pregnancy it is only about an eightieth part of that in the third week. In the still earlier stages it seems valid to assume that the relative rate of growth increases at the same proportionate rate as in the later ones; and, arguing from the more or less known volumes of the ovum at the time of conception, and of the embryo in its third week of growth, it has been roughly calculated that the first doubling in diameter of the fertilised ovum requires less than three days, or about 1-200th part of the time in the last month of pregnancy.

With regard to post-embryonic growth rate, but little need be said. A large number of observations have been made on the length of children at birth, and the mean of all the values for both sexes is approximately 50.0 cm. From Bowditch's tables of the statures of Boston school children,* it is found that the mean

* Report of Massachusetts State Board of Health, p. 275, 1877.

height of girls and boys between their 5th and 6th years is 105.2 cm., or only just over double the birth length. The rate of growth during this period is therefore only about 1-280th part of that in the third week of embryonic growth. Between the 5th and 10th years it is only about 1-900th part, and between the 15th and 19th years only 1-2400th part. Compared with the calculated rate of growth during the first day of development, the rate during this last period is about 5000 times smaller.

As regards another mammal, the guinea-pig, a very complete series of observations has been made by Minot,* though unfortunately these extend only to the post-embryonic stages. Minot has calculated the daily percentage increase in weight of his animals, and he found that this diminished steadily from the fourth day after birth onwards. During the first week or two it was over 5 per cent. per diem; at the end of the second month about 1.3 per cent., and after a year about .1 per cent. Minot seems to have been the first observer to recognise the importance of calculating the relative rate of growth, as distinguished from the absolute increments of weight. He considers there can be little doubt that there is a progressive loss of growth power at least in all mammals, and perhaps in all living beings.

That the rate of growth in the earliest stages of development is enormously more rapid than in the later ones may therefore be regarded as proven. And it seems to me that it is probably a general rule, with doubtless some exceptions, that the effect of environment on the growth of an organism depends chiefly on

* J. Physiol., xii. p. 97.

the rate of growth at the time. To what extent, however, does a retardation or acceleration of growth produced at any particular period of development persist to the adult stage? It is quite conceivable that such an effect, produced in the earliest stages, may be entirely compensated for by a subsequent variation of the growth in a reverse direction, and leave no trace behind. On the other hand, it is conceivable that it may persist unchanged and undiminished. In all probability, however, it does neither the one or the other. The initial effect gradually diminishes with progress of development, but never entirely disappears. An effect, once produced, can never be wholly eradicated. Upon this point some of the data given by Minot are again most pertinent. He calculated the arithmetical mean error in the weights of more than a hundred guinea-pigs, which were bred under as normal and favourable conditions as possible, and weighed at short intervals from the day of birth up to the end of their second year. During the first fortnight of growth, the individual weights were on an average found to vary from the mean weight by ±19.0 per cent. in male guinea-pigs, and ±17.2 per cent. in females. Between the 70th and 140th days this variability had sunk to ±12.7 per cent. for males, and ±14.7 per cent. for females. During the 145th to 215th days, it fell still further to ±7.5 per cent. for males, and ±13.0 per cent. for females. At later periods it remained constant in females, but increased again somewhat in males. Roughly speaking, therefore, one may say that with progress in development from the earliest to the latest stages, the variability becomes nearly halved.

Quite recently, Professor Pearson * has shown that in man also the diminution in variability with growth is very marked. Working on measurements of 1000 male and 1000 female new-born babies, he obtained the following values for the weight and length:

	MEAN WEIGHT.	MEAN LENGTH.	PER CENT. VARIABILITY WEIGHT.	PER CENT. VARIABILITY LENGTH.
Females,	7.073 lbs.	20.124 in.	14.228	5.849
Males,	7.301 lbs.	20.503 in.	15.664	6.500

Here we see that the male infant at birth is slightly heavier and slightly longer than the female, and also that its relative variability (error of mean square expressed as a percentage on the mean value) is distinctly greater. On working out similar data for 160 female students (mostly from 19 to 25 years of age), and for 1000 male students of the same age, Pearson obtained the following values:

	MEAN WEIGHT.	MEAN LENGTH.	PER CENT. VARIABILITY WEIGHT.	PER CENT. VARIABILITY LENGTH.
Females,	125.605 lbs.	63.883 in.	11.170	3.696
Males,	152.784 lbs.	68.863 in.	10.830	3.662

The woman is now slightly more variable than the man, both in weight and length, but the variability of both is very much less than at birth. It is, in fact, on an average, 26.4 per cent. less as regards weight, and 40.4 per cent. less as regards length.

As variability undergoes such marked diminution with age both in man and the guinea-pig, we may venture to agree with Minot that probably this diminution occurs in the growth of all mammals, and in fact that

* Proc. Roy. Soc., lxvi. p. 23, 1900.

variability of all characteristics may similarly diminish. Should this contention be justified by subsequent research, as seems very probable, then it might be formulated as a definite law. This could be worded as follows: "The variability of a developing organism diminishes regularly with its growth." Doubtless there are many exceptions or partial exceptions to this law. For instance, Bowditch * has shown that in the case of human stature, there is a distinct increase of variability just before, and at the time of, puberty in boys, and a slight one in girls, followed by a decrease. Weldon † has found that the variability of the frontal breadths of young crabs somewhat increases with growth, and then diminishes. Still these exceptions are not sufficient to upset the general validity of the law. As additional evidence in favour of it, may be cited the observations of Bumpus ‡ on the variability of the periwinkle. Calculating from measurements made on 13,000 shells from different sources, it is found that the variability in the ventricosity or relative breadth of the small and medium-sized shells is distinctly greater than that of the larger (and therefore on an average older) shells. Again, I myself have found that the variability in the size of *Strongylocentrotus* larvæ diminishes steadily from the fifth day of growth onwards.§

If our contentions as to rate of growth and variability be admitted, it follows that the variability of embryonic

* Report of Massachusetts State Board of Health, p. 275, 1877.

† Proc. Roy. Soc., lvii. p. 367.

‡ Zoöl. Bulletin, vol. i. p. 247.

§ Phil. Trans. 1895, B. p. 617 (and confirmed by subsequent observations).

animals should be very great indeed. Upon this point there is, unfortunately, very little evidence, though the general opinion of embryologists supports the view. However Fischel * has made several different measurements on 104 embryos of the duck, and he classified his results according to the number of vertebræ they contained. He found that the variability was very great, and also that it diminished considerably with growth, but the values are too irregular for brief quotation.

As the variability of guinea-pigs becomes nearly halved during post-embryonic development, it follows that any retardation or acceleration of growth produced in the earliest stages must also become nearly halved, or perhaps still more diminished. A certain amount of variation must have been produced in the growth of the animals by variation in their environmental conditions during development, though probably, as these conditions appear to have been kept as constant and as favourable as possible, this was not very large. One may perhaps assume, therefore, that if absolutely constant environmental conditions had been maintained, the variability, when growth had ceased, would not have been reduced to very much less than half its original amount. Now by the end of the first year the rate of growth had diminished to about a fiftieth its amount at birth, and hence, in order to produce equal permanent effects upon the growth, the period of exposure of an animal to a changed environmental condition shortly after birth would probably need to be only about a twenty-fifth part of the time required a year later. Again, to revert to the data ob-

* Morph. Jahrb., xxiv. p. 369.

tained in the case of man, we saw that between the 15th and 19th years the rate of growth was only about 1-2400th part of that obtaining in the third week of fœtal existence. Now granted that any effect produced by a changed environmental condition at this early period dwindled down to even a tenth of its original amount by the time the adult stage had been reached, yet the organism, as far as the permanent effect of environment is concerned, would still be 240 times more sensitive in the one case than in the other.

Striking evidence in support of this conclusion that retardations of growth, once produced, are never totally eradicated, has been recorded by Galton.* A friend of his took monthly measurements of the circumference of his children's heads during the first few years of their lives, and by plotting them out on paper, obtained curves of growth. These curves were regular on the whole, but each of them showed occasional halts, which were found to coincide in point of time with the various infantile diseases the children had experienced. These illnesses had therefore arrested the development, and this "had not been made up for by after growth." In fact "disease had drawn largely upon the capital, and not only on the income of their constitutions."

A few of Minot's data afford similar evidence. In one of his guinea-pigs, No. 34, the increase in weight was quite normal at first, but then absolutely ceased from the 90th to the 110th day. After this it continued normally as before, but the animal never succeeded in recovering its loss, and after two years weighed only about two-thirds as much as the other

* "Inquiries into Human Faculty," p. 235.

animals. One or two of the other observations show a somewhat similar, though not nearly so marked, result.

If our theoretical proof of the extreme sensitiveness of an organism to its environment during the earliest stages of development is held to be as valid for Mammals as the experimental proof showed it to be for Echinoderms, then we are met with a fact of some practical importance. We are afforded a scientific support of the widely held belief in the special influence of the condition of the mother at the time of conception upon the subsequently developing offspring. Should such conception occur when the blood and tissues of the mother are tainted by products of metabolism, the result of excessive fatigue, or by alcohol, or by the products of disease, there seems no reason why a considerable and lasting adverse influence should not be produced on the growth of the offspring. Still again, we must conclude that the health of the mother during pregnancy, and of the offspring during their early years, is of much more importance in its permanent effects on the physique and constitution than the health in later years.

Effect of Other Conditions on Variability. We have seen that the variability of a developing organism is far from remaining constant, but what is the condition of affairs in an adult organism? At the present day it has generally been more or less tacitly assumed that as a rule variability remains practically constant, and that external conditions, though they may affect the average values of the characters of an organism, do not influence the range of variation of these characters around their mean. In his Presidential Address, already referred to, Mr. Sedgwick has, however, come to a somewhat novel conclusion as to the change of variability with evolution. He points out that through the action of Natural Selection certain variations will be gradually eliminated, and the organisms will become more and more closely adapted to their surroundings. The variability of the species will therefore diminish. Thus breeders have found that "continued selection tends to produce a greater and greater purity of stock, so that if selective breeding is carried too far, variation almost entirely ceases." It follows, therefore, that "variation must have been much greater in past times than it is now. In fact, it must have been progressively greater the farther we go back from the present time."

This specious and apparently straightforward argument cannot be accepted, however, as it is built up on false premises. It is by no means true that by selective breeding we can reduce variability almost to nil. Professor Pearson* has calculated that, in the human race, if the parents be selected, then the variability of the offspring will be, on an average, only 9.5 per cent. less than that of the whole race; whilst if the grandparents and any number of more remote ancestors be selected as well, the variability will never become more than 10.56 per cent. smaller than that of the race. In fact, the variability of offspring, as compared with that of parents, depends directly upon the correlation between them. If the coefficient of correlation be r, then the variability of the offspring of selected parents will be

* "Grammar of Science," p. 485.

$\sqrt{1-r^2}$.* For instance, we have seen that in the case of man—and likewise also in other organisms—the coefficient of correlation between parent and offspring is .3, and between mid-parent and offspring .424. The variability of all the children of parents of any given stature will therefore be $\sqrt{1-.90}=.9539$, or 95 per cent. of that of the race, and the variability of the children of mid-parents of any given stature will be 90.55 per cent. of that of the race. Again, Professor Pearson and his co-workers † have found that in the series of measurements upon plants examined by them, there is no relation whatever between the variability of the species and the intensity of *homotyposis.* Thus on splitting up all the 22 series of measurements into two groups, the mean variability of the first group (as expressed by the coefficient of variation) was 22.95 per cent., and of the second, 14.28 per cent., or not two-thirds as much. Nevertheless the mean homotypic correlation (or degree of similarity between the undifferentiated like organs of an individual, as compared with organs of other individuals of the same race) was .456 in the first group, and .458 in the second; *i. e.*, was practically the same. "Hence there seems," says Professor Pearson, "so far as our researches go, no ground for asserting that increased intensity of heredity means decreased intensity of variation, and *vice versa.*"

To return to Mr. Sedgwick's views, if rigid artificial selection can only reduce variability by 11 per cent., then obviously Natural Selection can scarcely have any appreciable influence upon it. A more potent factor in diminishing variability, or at least in keeping it from increasing, is probably to be found in the greater fertility of the most commonly occurring forms of a race. As already mentioned in Chapter III., this phenomenon has thus far been shown to exist only in certain plants, and in a Hydromedusa, but the presumption is that it is of much wider application. Our knowledge of the matter is, however, insufficient to warrant an extended discussion.

An apparent instance of increase of variability accompanying cessation of the action of Natural Selection has been furnished by Bumpus,* in the case of the common sparrow. These birds were first introduced from Europe into America in 1850, and they have spread so rapidly, owing to their having abundant food and practically no natural enemies, that the continent is now inundated with them. The Natural Selection, which has kept the birds when in their native habitat more or less subservient to the regulations imposed by competing life, seems in this case to be largely suspended. According to Bumpus nearly all the young birds reach maturity; variations in colour and structure, unless most extreme, apparently not being disadvantageous to their possessor. In order to compare the variability of this American race of sparrows with the native European one, it was found convenient to make measurements of the eggs, rather than of the birds themselves, as they are so much less readily procurable. Eight hundred and sixty-eight American and 868 English eggs were compared with respect to length,

* Biol. Lectures, Wood's Holl, 1896–97, *also* Science, v. p. 423.

breadth, general shape, and colour. The American eggs were slightly more variable in length, they varying from 18 to 26 mm. instead of from 18.5 to 25 mm., and also were reduced in their average length from 22 to 21 mm. As Davenport has shown,* however, the mean deviation or arithmetical mean error, which is a better index of the variability, is somewhat smaller in the American eggs than in the English; but the arithmetical error of the variations in the relation of breadth of egg to length is slightly greater in the American eggs than in the English (in the proportion of .73 to .70). In order to compare the variations in the shape of the eggs, such as conical, ellipsoidal, pear and lemon shapes, Bumpus requested a disinterested person to select from the 1736 mixed eggs the hundred eggs which appeared to him most variable. The American eggs were known by a secret mark, and it was found that 81 of the selected eggs were American, and only 19 English. By a somewhat similar process of selection with reference to extremes of colour marking, it was found that 82 of the 100 selected eggs were American, and 18 English. There can be little doubt, therefore, that American eggs are considerably more variable, both in shape and colour, than English eggs.

The conclusions to be drawn from this most interesting series of observations, however, are probably not those suggested by the author. In the first place, as Davenport has pointed out, they afford presumptive evidence that the American eggs, or the birds which produced them, were subjected to a distinct selective process. Thus the curve of distribution of the length

* L'Année Biologique, 1897, p. 496.

of these eggs is distinctly asymmetrical, due, one may imagine, to the weeding out of a larger proportion of the birds producing the smaller eggs. The curve of distribution of the English eggs is practically symmetrical, so that selection in them probably removed extreme individuals to an equal extent in either direction. It should be borne in mind, however, that the skewness of the curve of distribution of the American eggs may, after all, be due only to Reproductive Selection, *i. e.*, to a greater fertility of the birds producing the larger eggs.

To what, then, are the diminished size of the eggs and their increased variability due? Probably they are nothing more than a direct response to the change in the environmental conditions. These conditions, though, by virtue of the absence of enemies and other checks, more suitable to increase in the numbers of the sparrows, are probably more variable, and, on the whole, not so suitable to individual growth. Thus in England the sparrows have been subjected to a very keen competition for a very long time, and so have become closely adapted to their surroundings. In other words, their environment is as favourable as it can be for their maximum growth, and for a minimum of variability.

Another series of observations made by Bumpus * to prove his views may from their intrinsic interest be cited also. They concern the variations of the periwinkle, *Littorina littorea.* This mollusc appears to have been first carried to American shores about 1850, and since then it has spread with great rapidity along the New England coast. For statistical purposes, 1000

* Zoöl. Bulletin, vol. i. p. 247.

specimens of the shell were obtained from each of ten different localities in America, and from three in Great Britain. The variation was estimated by determining the ratio of breadth to length in each shell, or the ventricosity, and expressing the result as a percentage of the one on the other. The extreme variations recorded were respectively 80 per cent. and 104 per cent., or a difference of 24. As a measure of the variability Bumpus took the difference in the extreme values exhibited by each sample of shells, and he found that the amplitude of variation of the American shells was in every case in excess of that of the British shells. Thus it was respectively 18, 19, 20, 19, 17, 20, 20, 18, 18, and 20 in the American samples, and 15, 14, and 12 in the British. Obviously, however, this is not a very good method of estimating variability, as it is more or less a matter of chance if specimens of shells exhibiting the most extreme variations happen to occur in any given sample of 1000. Thus in two-thirds of all of Bumpus' series, the extreme variations were represented by only a single specimen. Fortunately the true value of these extensive series of observations has not been lost, as Duncker * has taken the trouble to work through them and calculate the variability by an exact method. The following are the values obtained by him for the mean ventricosity, and its variability (as expressed by the error of mean square) in each sample.

From the table we gather that the average ventricosity of the American shells is slightly greater than that of the British (91.01 as against 89.78, or 1.4 per

* Biol. Centralblatt., vol. xviii. p. 569, 1898.

cent. greater). The variability is in every case considerably greater for the American shells, as Bumpus had concluded. The average variability of all the American samples is 2.765, or 18.2 per cent. larger than the average for the British (2.34).

	LOCALITY.	MEAN.	VARIABILITY.
I.	Tenby, Wales,	90.96	2.38
II.	South Kincardineshire, Scotland,	87.85	2.34
III.	Humber District,	90.53	2.30
IV.	St. Croix River, Maine,	91.26	2.70
V.	Casco Bay, Maine,	92.53	2.67
VI.	Beverly, Mass.,	90.65	2.76
VII.	Nahant, Mass.,	92.19	3.03
VIII.	Plymouth, Mass.,	90.09	2.48
IX.	Seaconnet, R. I.,	89.72	2.86
X.	Newport, R. I.,	89.17	2.62
XI.	Bristol, R. I. (shingle),	90.77	2.75
XII.	Bristol, R. I. (sand),	91.07	2.83
XIII.	Warren River, R. I.	92.69	2.95

Bumpus found that in other respects also the American shells are more variable than the British. Thus they show greater variations of ventricosity in the course of growth; shells of a given length exhibit a greater range of variation in weight; also their colour markings are more variable. Hence there can be no doubt that the general variability of the organism has increased. As in the case of the sparrow's egg, Bumpus attributes this to its emancipation from many of the restraining influences of Natural Selection; but it seems to me a simpler explanation, as was suggested in the case of the eggs, to regard the effects produced as largely, if not entirely, the direct result of want of adaptation to the changed and perhaps more variable conditions of life.

That want of adaptation should, as a rule, lead to increased variability may not seem at first sight obvious, but a little reflection will show, I think, that it is only what one ought on *a priori* grounds to expect. Thus supposing a group of organisms is subjected to unfavourable environmental conditions, so that they are all decreased in size by, on an average, say 10 per cent. We know by experience that all organisms are not equally affected by changed conditions. Some of them are more resistant than the average, and others less. Consequently, supposing the group of organisms had originally varied between such and such limits, then on reduction of their average size by this 10 per cent., some of the small individuals, and likewise some of the large ones, will be diminished by perhaps 12 per cent. in size, and others by only 8 per cent., and hence the range of variation of the whole group of organisms will be widened.

Actual proof of this contention has been obtained by the author in the case of sea-urchin larvæ. Thus, in experiments of one kind and another, some 412 series of measurements, each on 50 individuals, have been made on the body length of *Strongylocentrotus* larvæ after 6 or 8 days' growth under various favourable and unfavourable conditions. The variability of each series was determined by a method it is unnecessary to mention in detail here,* that of the whole of the larvæ being taken as 100. All the series of larvæ grown under so-called "normal" conditions (*i. e.*, in jars of sea-water kept at as constant a temperature as possible) and of other larvæ which did not vary from the normal by more than

* *Vide* Phil. Trans., 1895, B. p. 617.

±1.9 per cent., were put into one group, and a mean taken of their "variability" numbers. The larvæ which, owing to a more favourable environment, were larger than the normal, and which differed from it by from + 2 to + 4.9 per cent., were similarly grouped together. Still other groups were formed of those differing by from + 5 to + 9.9 per cent. and by + 10 per cent. and more. Like groups were made of larvæ which adverse environmental conditions had rendered smaller than the normal. The following are the mean results:

VARIATION IN BODY LENGTH OF LARVÆ FROM NORMAL.	NUMBER OF SERIES MEASUREMENTS.	VARIABILITY
+ 10 per cent. and over, ...	26	89.6
+ 5 to + 9.9 per cent., ...	30	96 8
+ 2 to + 4.9 per cent., ...	51	96.4
± 1.9 per cent., ...	170	96.8
− 2 to − 4.9 per cent., ...	63	100.8
− 5 to − 9.9 per cent., ...	43	104.4
− 10 per cent. and over, ...	29	113.2

Here we see that, with one slight exception, the variability becomes steadily greater and greater as the environment becomes more and more unfavourable, the extreme value in one direction being 26.3 per cent. greater than that in the other. This experimental result, bearing out as it does our theoretical conclusion, seems to justify one in assuming the existence of a definite law of variability. This may be worded as follows: "An organism varies least when it is best adapted to its surroundings, so that the less it is adapted, the more variable does it become." Extended research alone can teach us how general may be the application of this law, and afford adequate numerical estimates as to the relation between want of adaptation and variability in different organisms. Without doubt, however, these must vary enormously.

It may be thought that this law is a mere reproduction, in different words, of one of the numerous causes of variability discussed by Darwin. I do not think that this is the case, however, for though he states that variability of every kind is due to changed conditions of life, such as climate, food, and excess of nutriment, he does not suggest that the real factor is want of adaptation between the organisms and their surroundings. On the contrary, he endeavours to show that the amount of modification which animals and plants have undergone does not correspond with the degree to which they have been subjected to changed circumstances.

From the conclusions at which we have arrived, it follows that increased variability of environment leads to increased variability of the organisms subjected to it. A proof that this is the case has been afforded by Montgomery,* who has studied the variation in migratory and non-migratory species of North American birds. From data collected by Ridgway in his "Manual of North American Birds," he calculated the percentage amount of variation (according as it was under 1 per cent., between 1 and 1.5 per cent., between 1.5 and 2 per cent., or over 2 per cent.) in the culmen of the bill, wing, tarsus, and tail of the species and sub-species of fifty-six families. His results show "that migratory species evince a greater amount of individual variation than do non-migratory species; and species which undertake extensive migrations, a greater amount than

* J. Morph., 1896, p. 25.

species which make migrations of less magnitude." This is in conformity with the rule that species inhabiting more extensive and diversified areas show more individual variation than those with small or insular breeding areas, and also with another rule, "that species with geographical races evince a greater amount of individual variation than do species which are not divided into such races, provided that the breeding area is approximately equal in extent and diversification in both cases." Again, it was found that males generally exhibit a greater amount of individual variation than females of the same species, the males being most variable in 60.4 per cent. of the 273 cases investigated, and the females most variable in 39.6 per cent. of them.

From these facts Montgomery deduces what he considers to be a very important conclusion, viz., that "individual variation is greater in amount in those species which we must consider under the influence of a continuous process of development than in those species which we must consider as being influenced by no process of development at all." Though this view seems highly probable, yet it appears to me that the data adduced do not in any way prove it. Because a bird occupies an extensive breeding area, or is migratory, it does not follow that it is undergoing development either progressive or retrogressive. The only deduction which seems to me permissible is the one above mentioned, one which has, indeed, been drawn by Montgomery himself in the following words: "The amount of individual variation stands in a direct ratio to the degree of complexity of the environmental forces which influence the organism."

One of the few conditions which have been generally held to lead to increased variability is that of *Domestication.* Darwin * laid considerable stress on the fact that "with extremely few exceptions all animals and plants which have been long domesticated have varied greatly." This increased variability he attributed partly to the conditions of life being less uniform, and to a lesser degree to the effects of excess of food. He concluded that organic beings must be exposed to the new conditions for several generations before much increase of variability is observed, and hence it seemed as if the changed conditions of life acted not only directly on the organism as a whole, but also indirectly through the reproductive system.

In the first place it is probable that Darwin considerably overestimated the variability of domesticated organisms, as compared with that of undomesticated ones. Thus we have already seen how widely many naturally occurring organisms may vary, and Schwalbe has even denied altogether that domestication produces any increase of variation. Thus he states † that Pfitzner has found that parts of the skeleton of the fore and hind feet of wild animals vary just as much as the corresponding parts in man. The size and the indices of the skull of the otter, as determined by examining over 200 skulls derived from a limited district in Alsace, were found by Schwalbe to vary just as much as the corresponding measurements in the most widely separated human races. Again, Kohlbrügge has shown that in all

* "Animals and Plants," ii. p. 401.

† Anatomischer Anzeiger. Verhandlungen, xii. p. 2, 1898.

the species of Primates examined, the muscles were quite as variable as in man.

Though Schwalbe's views may be correct as regards some domesticated races, they obviously cannot apply to all. For instance, domesticated pigeons, dogs, and horses vary between very much wider limits than any naturally occurring species. But it is obvious that such extreme variations as these owe their origin chiefly to careful artificial selection, extending over very large numbers of generations. When no such selection is practised, as in some of the domesticated animals kept by savages, then it is highly probable that the variability is no greater than in allied feral species. Any slight increase of variability actually present may doubtless be attributed to the direct action of the changed conditions of life.

Darwin's conclusion that increased variability only shows itself after several generations of domestication, if true as it stands, is difficult to account for in any other way than by supposing that the reproductive system is affected. But it seems to me distinctly probable that much, if not all, of the increased variability may be ascribed merely to the careful selection of the most aberrant individuals in each generation, and their assortive mating. In such a case the variability of the whole group would obviously increase from generation to generation.

CHAPTER VII.

THE EFFECT OF TEMPERATURE AND OF LIGHT.

Variations and modifications—Effect of temperature on growth of frog—Optimum temperature of growth in plants—Effect of temperature on size of sea-urchin larvæ, of Lepidoptera, and of Mollusca—Seasonal dimorphism in certain Lepidoptera in its relation to temperature—Temperature differences giving rise also to local races, sports, and phylogenetic forms—Critical period of reaction to temperature—Effect of Arctic climate on coat of mammals—Effect of darkness and of light on growth of plants—Effect of sunlight and of diffused light—How far does pigmentation of animals depend on exposure to light?—Cave animals—Illumination of under surface of flounder—Effect of light and of darkness on Molluscs—Variable protective resemblance in the frog, in fish, and in larvæ and pupæ of certain Lepidoptera.

HAVING discussed blastogenic or germinal variations as fully as our very limited knowledge of the subject permits, we will now proceed to enquire into the other great class of variations, namely, those produced by the action of the environmental conditions on the soma or body tissues. These were termed by Weismann somatogenic variations, but Lloyd Morgan, Mark Baldwin, and others, with a view to distinguishing them more pointedly from the former class of variations, have referred to them as *somatic modifications*, or more simply as *modifications*. The term "variations" they reserve exclusively for blastogenic variations. Though doubtless there is much to be said for this system of classifying variations, yet there is also something against it. It is frequently difficult or impossible to decide whether any given variation is of blastogenic or somatogenic origin; frequently, indeed, it is partly the one and partly the other. Hence the term "variation" is convenient for general use, when one wishes to imply nothing as to the origin of the observed condition, whilst the narrower meaning may be applied to it when it is mentioned in connection with the recognised and closely defined term "modification."

In our discussion of somatic modifications, it is found most convenient to classify them according to the agencies which have produced them, and not according to the effects produced. Those observations in which there appears to be a clear and direct relation between cause and effect, between particular environment and particular modification, will first be mentioned, and the mixed and often indirect effects of more complex conditions, such as climate and general environment, studied later.

Temperature. Of all the environmental conditions, temperature is probably one of the most unequivocal in its direct production of variations. That is to say, in the majority of cases it acts directly on all the cells composing the tissues, promoting or retarding their growth, and producing a permanent effect on the organism as a whole. Of course it does not necessarily follow that, because the rate of growth is altered, a permanent effect is produced on the absolute size of the organism. Yet in the light of the observations described in the last chapter, we are justified in assuming that such is as a rule the case in animals. For plants it will probably be generally admitted, if the difference in the size of trees and plants grown in the tropics and in temperate climates, or in temperate and arctic ones, be borne in mind.

Some striking observations on the effects of temperature on growth were made by Higginbottom some fifty years ago.* Ova of the frog were allowed to develop under otherwise similar conditions at the temperatures 15.6°, 13.3°, 11.7°, and 10.5° C. The tadpoles were observed to leave the ova respectively 9, 14, 20, and 20 days after the beginning of the experiment, whilst frogs were fully developed after respectively 72, 160, 170, and 234 days. Thus a difference of 5° C. more than trebled the period of development.

AMBLYSTOMA TIGRINUM.			RANA VIRESCENS.		
Average Temperature.	Average Time of Development.	Number of Observations.	Average Temperature.	Average Time of Development.	Number of Observations.
4°C.	288 hours	2	4°C.	471 hours	1
8°	210	4	8.75°	192	1
9.5°	139.2	5	12.12°	126	2
13°	96	2	16°	60	3
14°	90	2	22°	27.5	2
18°	60	3	24°	25.5	2
22°	40	6	26°	21.5	1

Within the last few years, a considerable number of observations have been made. Thus Lillie and Knowlton † experimented with the ova of *Amblystoma tigrinum* and of the frog *Rana virescens.* The time of development from the first or second cleavage to the last

* Phil. Trans. 1850, p. 431.

† Zoöl. Bulletin, vol. i. p. 179.

stage of disappearance of the yolk plug was determined, these being the most sharply marked periods. The temperatures varied between 4° and 26° C., several observations being made at each temperature, and means taken.

From the above table we may gather that at the

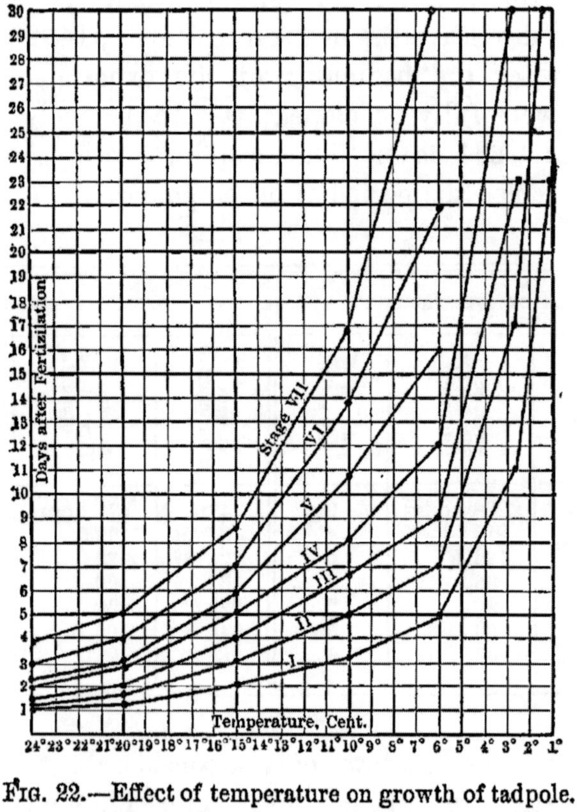

FIG. 22.—Effect of temperature on growth of tadpole.

highest temperatures employed the rate of development was respectively 7.2 and 21.9 times more rapid than at the lowest.

In the accompanying figure are reproduced the results obtained by O. Hertwig * upon the ova of *Rana fusca.* The curves represent the number of days required by the ova, kept at different temperatures, to reach certain definite stages. The ordinates indicate the time in days after fertilisation, and the abscissæ the temperature. For the lowest curve the stage to be reached was that of a gastrula with the blastopore closing in, and we gather from this curve that the time required at a temperature of 1° C. was 23 days; at a temperature of 6,° 4.9 days, and at a temperature of 24°, only a single day. Stage II was that of an embryo having a rudimentary medullary plate, with its edges rising and separated by a broad cleft; Stage III that of an embryo with a closed medullary tube, and with a distinctly marked head: Stage IV that of a more elongated embryo with an obvious tail, but with gills not formed; Stage V that of an embryo 5 mm. long, with rudiments of gills; Stage VI that of an embryo 7.5 mm. long, with well-developed gill tufts and tail 3.5 mm. long; Stage VII that of an embryo 9 mm. long, with a tail 5 mm. long. The curves representing the times of growth to all these more advanced stages are very similar to each other and to the first curve.

In the observations thus far quoted, the highest temperature employed was 26°, and this proved to be also the most favourable temperature for growth. Higher temperatures still may produce an adverse influence, as we have already observed in the case of Echinoid larvæ. For the growth of tadpoles' tails, Lillie and Knowlton found the "optimum" temperature to be 30°, the rate of increase in length then being 10.6 times greater than

* Arch. f. mik. Anat., li. p. 319, 1898.

at 10°. At 31° to 34.9°, however, the rate was only 9.0 times greater. Better instances of the more and more unfavourable influence of increasing high temperature are found amongst plants, as in them the optimum temperature is much further removed from the "maximum" temperature (the highest temperature at which growth can take place at all) than it is in animals. The following table shows the increments in the length of the hypocotyls of various plants in a period of 48 hours, as determined by Koppen and by De Vries:*

	KOPPEN.			DE VRIES.		
Temperature.	Lupinus albus.	Pisum sativum.	Zea mais.	Sinapis alba.	Lepidium sativum.	Linum usitatissimum.
14.1°C.	9.1 mm.	5.0 mm.				
15.1				3.8 mm.	5.9 mm.	1.5 mm.
18.0	11.6	8.3	1.1 mm.			
21.6				24.9	38.0	20.5
23.5	31.0	30.0	10.8			
26.6	**54.1**	**53.9**	29.6			
27.4				**52.0**	**71.9**	**44.8**
28.5	50.1	40.4	26.5			
30.2	43.8	38.5	64.6			
30.6				44.1	44.6	39.9
33.5	14.2	23	**69.5**			
33.9				30.2	26.9	28.1
36.5	12.6	8.7	20.7			
37.2				10.0	0.0	9.2

The optimum temperature was about 27° for every plant but one, viz., *Zea mais*, and in this case it was 33.5°. The rate of growth at the optimum was, in the various plants, respectively 5.9, 10.8, 63.2, 13.7, 12.2, and 29.9 times greater than at the lowest temperature at which it was observed, and respectively 4.3, 6.2, 3.4,

* Quoted from Vines' "Physiology of Plants," p. 293.

5.2, and 4.9 times greater than at the highest temperature. In one plant, however, the temperature of 37.2° was sufficient to stop all growth, so that this was a slightly supra-maximal temperature.

That the increased growth produced by warmth leads, at least in some cases, to an actual increase in the size of an organism, is proved by some further observations of the author on Echinoid larvæ. No special determinations were made as to the effect of temperature on growth, but it was noticed that, whilst at a temperature of 13.8°, the ova took about 22 hours to reach the free swimming blastula stage, they took only about 5 hours at 24°. We may assume, then, that the rate of growth increases rapidly with temperature. The effect of various temperatures on the actual size of the larvæ may be gathered from the following table:*

Temperature of Development.	Strongylocentrotus.		Sphærechinus.		Echinus microtuberculatus.	
	Body length.	Arm length.	Body length.	Arm length.	Body length.	Arm length.
11.4°	100.0	100.0	100.0	100.0	100.0	100.0
15.9°	113.5	143.4	109.4	287.0	113.4	116.3
20.4°	120.6	156.8	104.6	327.2	124.5	106.6
23.7°	122.5	149.1	100.6	386.7	123.9	113.7

Larvæ of three different species were allowed to develop at four different temperatures, and measured after 8 days' growth in respect of both their body length and their anal arm length, and this latter measurement was calculated as a percentage on the former. In the case

* Phil. Trans. 1898, B. p. 479.

of both *Strongylocentrotus* and *Echinus* plutei, the body length increases considerably with the temperature up to 20.4°, when it is more than 20 per cent. greater than in plutei grown at 11.4°, but it is practically unaffected by a further rise. In *Sphærechinus*, on the other hand, the optimum temperature appears to be 15.9°, and a further rise of temperature acts unfavourably. The effects on the arm lengths differ considerably more than those on the body lengths. In *Strongylocentrotus* this dimension is half as long again at 20.4° as it is at 11.4°, but in *Echinus* it is only very little affected. In the *Sphærechinus* pluteus, on the other hand, it is nearly four times longer at 23.7° than it is at 11.4°. Each organism, therefore, in respect of each portion measured, reacts in a different manner to changes in the temperature of development.

Some observations of Standfuss * upon the larvæ of certain Lepidoptera, show that the effect of temperature on growth is not necessarily in the direction of increased size. Thus he found that when, as was often the case, the larval period was shortened by raising the temperature, the size of the adult insects resulting therefrom was correspondingly reduced. For example, a pair of *A. fasciata*, of which the wings measured respectively 46 and 48 mm. across, produced three specimens measuring only 36 to 39 mm., when the larval stage was reduced to 68 to 87 days, and the pupal to 15 to 20 days, by subjection to a temperature of 25° to 30°. On the other hand, some other eggs from the same original pair of *A. fasciata*, which, though exposed

* The Entomologist, vol., xxviii. p. 69, 1895. (Translated from the German by Dr. F. A. Dixey.)

to the same high temperature, developed more slowly—the larval period taking 142 to 163 days, and the pupal 25 to 31 days—yielded specimens having a wing measurement of 55 to 57 mm. It would seem, therefore, that high temperature may so hurry forward the time of onset of the metamorphosis from larva to pupa, that there is insufficient opportunity for adequate larval feeding and growth, and a consequent dwarfing of the adult imago. If, however, there be no curtailment of the normal period of feeding, the high temperature may produce a considerable increase in the size of the individuals. These conclusions are supported by several other observations.

The permanent effects of temperature on size are probably very considerable among many of the Mollusca. Thus it has often been noticed that snails living in cold and exposed positions are considerably dwarfed in comparison with those living in warmer regions, but, as far as I am aware, no exact comparisons have been made. Even if this had been the case, it would not be permissible to ascribe the differences to the direct result of temperature, as this might have acted indirectly, through the vegetation. Certain observations of Möbius * on marine Mollusca seem, however, to demonstrate the direct effects of temperature changes. Thus it was noticed that the Molluscs in the Eastern basin of the Baltic are much more stunted than those in the Western. For instance, *Mytilus edulis* is only 3 to 4 cm. long near Gothland, whereas at Kiel it attains a length of 8 to 9 cm. Also in the Eastern basin the calcareous layers of certain shells such as *Mya arenaria*

* Report on "Pomerania" Expedition, p. 138.

are extremely thin. "These remarkable variations are, no doubt, to a large extent due to the violent changes of temperature which are experienced in the Baltic, and by which the steady development of the animals in question is thrown out of gear. The same species occur on the coast of Greenland and Iceland, where they attain a considerably larger size than in the Baltic, in spite of the lower mean temperature, probably because their development is not interrupted by any sudden change from cold to heat, or *vice-versa.*"*

The influence of a low temperature on the colour of marine Mollusca seems to be indicated by the observations of Fischer † on the shells of the west coast of South America. Numerous species of these shells exhibit a remarkable degree of melanism, and it seems highly probable that "this concurrence of specific melanism (which stands quite alone in the world) is due to the cold polar current which impinges on the Chilian coasts, for the same genera occur on the opposite shores of the continent without exhibiting any trace of this mournful characteristic." ‡ It is very improbable, however, that this melanism is the direct result of the cold current. If so, why should it not be observed in other parts of the world, which are similarly visited by cold currents?

More interesting and unequivocal effects of temperature are afforded by the numerous experiments which have been made upon the wing colours and markings of

* Quoted from Cooke, "Cambridge Natural History," vol. iii. p. 84.

† Journ. de Conchyl., xxiii. p. 105, 1875.

‡ *L. c.*, p. 85

Lepidoptera. It has been known for more than sixty years that the two butterflies *Vanessa levana* and *V. prorsa*, formerly regarded as different species, are but seasonal forms of one and the same species. Thus *V. levana* emerges in the spring, breeds immediately, and produces adult *V. prorsa* progeny in the same summer. The progeny of these insects pass the winter as chrysalids, and emerge the next spring as *V. levana*. The *levana* form is characterised by a yellow and black pattern on the upper side of the wings, whilst the *prorsa* form has black wings with a broad white transverse band. The lower surfaces differ only slightly.

It is a natural supposition that these changes of colour marking are dependent in some way on temperature, and Dorfmeister * proved that this is actually the case. By the application of warmth to the pupæ he succeeded in producing *prorsa* out of the offspring of *prorsa*, and by the application of cold he obtained from *levana* not the pure *levana* form, but one intermediate between it and *prorsa*. This intermediate form, which has occasionally been observed in nature, is termed *V. porima*. These experiments were repeated and extended by Weismann, and by employing a greater degree of cold he succeeded in obtaining *levana* from *levana;* but he found that *prorsa* was only exceptionally reared from *prorsa* by the application of heat.

The mode of action of the temperature is not so clear as might at first sight be imagined. The simplest explanation is to attribute the effect to the direct influence of the warmth and cold, and this view of the in-

* Mitt. des naturwiss. Vereins für Steiermark, 1864. See also Eimer's "Organic Evolution," English Ed., p. 116, *et seq.*

fluence of warmth is actually held by Eimer.* According to Weismann, however, the action, both for warmth and cold, is an indirect one. The change of the progeny of *levana* back to *levana* through the influence of cold he attributes to reversion to the ancestral form, for there is practically no doubt that *levana* is phylogenetically the older form of the two. He considers that *prorsa* has slowly arisen through the gradual increase in the warmth of the climate, or is a seasonally adaptive form, and that its occasional production from the progeny of *prorsa* is due to the high temperature unduly stimulating the development of the *prorsa* "determinants."

A clearer case of the direct influence of warmth and cold is afforded by *Polyommatus phlœas*, the Small Copper butterfly. By exposure of the pupæ to various temperatures, Merrifield † obtained the following results:

TEMPERATURE.	TIME OF EMERGENCE.	COLORING OF SPECIMENS.
27°—32° C.	6 days	Spots large, not sharply defined; dusky suffusion of fore wings.
about 21°	11—15 days	Spots smaller; copper colour more vivid; black more intense.
about 14°	22—23 days	Copper colour still more vivid; copper band on hind wings broader.
about 7°	57—59 days	Effects intensified.
.5°; then 13°	10 weeks; then 5 wks.	Extreme effects, especially in smallness of spots and breadth of coppery band on hind wings.

Here we see that the temperatures ranged from about 30° C., or 85° F., to just above the freezing point. The

* "Organic Evolution," p. 122.

† Trans. Ent. Soc. 1893, p. 55.

times of emergence of the butterflies from the chrysalis varied from 6 days to no less than 15 weeks, and probably if the low temperature had been continued in this latter case, the time would have extended to many months, or there may have been no emergence at all. It will be seen that the principal effects produced by warmth are a dusky suffusion of the fore wings, and by cold an intensity of colouring in both the coppery and dark parts, the enlargement of the copper band on the hind wings being an especially marked feature. In fact these "cold" specimens were very similar to those caught in England, Germany, and similar latitudes, whereas the "warm" specimens were similar to the variety *eleus*, which is found in Southern Europe. Merrifield therefore came to the conclusion that the difference in the appearance of these local forms "is not necessarily to be attributed to the existence of races of different colouring, but may be owing to the difference between the temperatures to which the individuals are exposed in the two climates." Weismann has shown,* however, that the modifications cannot be entirely due to the direct effects of temperature. Thus none of the specimens obtained by exposing pupæ of a German stock to high temperature were so dusted with black as the darkest forms of the southern variety *eleus*, whilst conversely, none of the specimens obtained by exposing the pupæ of a Neapolitan stock to a low temperature were so light-coloured as the ordinary German form. "The German and Neapolitan forms are therefore constitutionally distinct, the former tending much more strongly towards a pure reddish-gold, and

* "Germ-Plasm," p. 399.

the latter towards a black colouration." Weismann thinks that the two varieties may have originated owing to a gradual cumulative influence of the climate, the slight effects of one summer or winter having been transmitted and added to from generation to generation. Such a cumulative effect can be accounted for satisfactorily by supposing that the temperature not only affects the "primary constituents" of the wings of each individual—*i. e.*, a part of the soma—but also the corresponding "determinants" of the germ-plasm contained in the germ cells of the animal.

Arguing from experiments on about 5000 pupæ, Standfuss * has endeavoured to classify under five different headings the effects which temperature changes may produce in Lepidoptera.

(1) They may give rise to seasonal forms having a similar aspect to those occurring among the palæarctic fauna at certain definite seasons of the year. For instance, pupæ of *Vanessa c-album* (Comma butterfly), kept at 37° C., gave origin to the light coloured, yellowish brown form of butterfly, especially pale on the under surface, whilst those kept in a refrigerator produced the form with a considerably darker under side, in many cases mingled with a moss-green tint. Also this form had much more sharply defined markings, and a more deeply indented margin to the wings. Both these forms, be it noticed, occur in nature at the present time. Again, by exposing pupæ of *P. machaon* (Swallow-tail), to a temperature of 37° C., insects were obtained which bore a perfect resemblance to those that

* The Entomologist, vol. xxviii. pp. 69, 102, and 145. (Translated from the German by Dr. F. A. Dixey.)

fly in August in the neighbourhood of Antioch and Jerusalem. Pupæ kept at 5° to 8°, however, yielded the common Swiss and German form of butterfly obtained from hibernated pupæ.

(2) Local forms and races such as occur constantly in certain definite localities may be produced. For instance, exposure of pupæ of *V. urticæ* (Small Tortoiseshell) to warmth produced specimens somewhat similar to the variety *ichnusa*, whilst cold produced some specimens which strongly recall the North American *V. milberti*, and others which were indistinguishable from the northern variety *polaris*. Again, warmth acting on pupæ of *V. cardui* (Painted Lady), gave an extraordinarily pale form, like those found in very different parts of the tropics. Cold, on the other hand, gave specimens with a very recognisable darkening of the whole insect, such as is exhibited by a form found in Lapland.

(3) There may arise forms of an entirely similar aspect to some which are also found exceptionally under natural conditions, *i. e.*, aberrations. For instance, warmth, acting for a brief period on *V. cardui*, produced a few specimens of the aberrant form *elymi*. Cold, acting on pupæ of *V. io* (Peacock), produced a variety ab. *fischeri*, which exhibits a reduction in the number of the blue scales on both fore and hind wings. In these and other characters an approach to the type of *V. urticæ* is perceived. Such observations as these suggest that a large number of the aberrations occurring in nature may have actually arisen through the influence of abnormal temperature conditions.

(4) There may be produced *phylogenetic* forms;

forms, that is, which are nowhere to be found on the earth at the present day, but which may have existed at past epochs. Such a result may have been effected through modification of temperature conditions having taken place in the actual habitat of the species, or from the species having migrated to a more southerly or northerly region. The variety *fischeri* of *V. io*, just mentioned, is probably a phylogenetic form. The same may be true of a variety *rœderi* of *V. antiopa* (Camberwell Beauty), which Standfuss obtained by keeping the pupæ in a refrigerator. Again, exposure of the pupæ of *V. atalanta* (Red Admiral) to warmth, produced specimens approximating towards *V. callirrhoë* and its local forms, such as var. *vulcanix*, which are found in the Canaries: *i. e.*, to forms which may resemble the common ancestor of these species. Other forms were produced which may perhaps be destined to arise in the future, in that they are further removed from the type of related species, instead of approximating to them, like the true phylogenetic forms. For instance, the widely diverging specimens obtained in a few instances by the action of warmth on *V. antiopa*, may belong to this class. This variety has been named *daubi* by Standfuss.

(5) Finally, there is still a small unexplained residue of modifications produced by temperature changes. This possibly represents the direct reaction of the individual species, completely independent of, and uncontrolled by, any inherited developmental tendency.

It will be noticed that the principle of reversion is called in by Standfuss to account for one of his five groups, but Weismann, Dixey, Fischer, and others are inclined to extend its scope to some of the other cases as well. Thus Weismann formerly made use of it to account for seasonal dimorphism, though now he rather withdraws this opinion.* According to Fischer,† both very low and very high temperatures are equally capable of determining reversion by acting simply as excitants. A moderate elevation of the temperature, on the contrary, may give rise to new modifications which are not phylogenetic, but which actually occur in warm climates. Dixey,‡ arguing especially from Merrifield's observations on *V. atalanta,* and Merrifield § himself, from these and other observations, have come to the conclusion that reversion may be occasioned by exposure both to high and to low temperatures, but that the kind of effect produced is different in the two cases. Eimer is of the opinion that only cold has the power of causing a reversion to an ancestral form, the effect of warmth being "evidently a direct effect." || In support of his views, he refers to Weismann's experiments on *Pieris napi* (Green-veined White), and *V. levana-prorsa.* The former butterfly occurs in a summer and a winter form, the winter being the darker. There is also a variety of *P. napi*, viz., *bryoniæ,* which is found in the Swiss Alps and in the polar regions, and which can be described as a very dark variety of the winter form of *P. napi.* This *bryoniæ* is in all probability the ancestral form of *P. napi,* whilst the winter form, and

* The Entomologist, 1896, p. 240.

† "Transmutationen der Schmetterlinge infolge Temperaturänderungen," Berlin, 1894.

‡ Trans. Ent. Soc. 1893, p. 72.

§ Trans. Ent. Soc. 1894, p. 425.

|| "Organic Evolution," p. 125.

subsequently the summer form, of the common *P. napi*, have probably arisen gradually from it through the influence of a warmer climate. Now Weismann found that he was unable to convert *bryoniæ* into *napi* by the action of warmth, though he could by the application of cold readily change the summer form of *napi* into the winter form. Similarly, also, the progeny of *V. levana* are readily converted by cold into *levana*, but only exceptionally can the progeny of *V. prorsa* be converted into *prorsa*. Now, as already mentioned, *levana* is probably the ancestral form, and so, in both this case and that of *P. napi*, cold readily produces what is probably a phylogenetically older form, whilst warmth generally has no effect. Certain observations by Merrifield * also afford some support to Eimer's view, for he found that "the capability of being turned during the pupal period from one type partially into the direction of the other exists in both the summer and the winter type, but is *much* greater in the former than in the latter."

With regard to the critical period at which temperature especially exerts its influence, there is a general consensus of opinion that it is confined to the pupal stage, and in most cases also to the first part of this stage. Dorfmeister † concluded that temperature exerted its greatest influence during the change from the larval into the pupal stage, or shortly afterwards. Weismann ‡ noticed that in *V. prorsa-levana* it acted only at the beginning of the pupal stage. Standfuss,

* Trans. Ent. Soc. 1892, p. 53.
† *Vide* Eimer's "Organic Evolution," p. 131.
‡ "Germ-Plasm," p. 402.

in almost all the observations above referred to, exposed his pupæ to warmth for about three days, and then kept them at the room temperature until they emerged, this generally occurring 4 to 10 days later. The exposure to cold generally extended to about 30 days, and emergence took place after about 11 days more at room temperature. As the effects obtained by him are just as great, if not greater, than those obtained by other observers, it would seem quite clear that in the forms he employed the critical time for temperature is certainly during the first portion of the pupal period. However, Merrifield,* in his observations on the summer and winter forms of *P. napi*, found the critical time to be in the last days of the pupal period, a directly opposite result to that of Weismann for the same insect. Weismann † explains the apparent contradiction by supposing that in *P. napi* adaptive and direct seasonal dimorphism are mixed. The species may have adapted itself to the seasons of the year by a double protective colouring, and the critical period for the determination of the adaptive form may be at the beginning of the pupal period. The direct reaction of the species to temperature may, however, as Merrifield found, be determined only at the end of the pupal period.

In his experiments with *P. phlæas*, Merrifield found that pupæ kept at 0.5° C. for ten weeks, and then exposed to a temperature of 32° for six days, gave specimens with features very similar to those obtained from pupæ kept throughout at a temperature of 27° to 32°.

* Trans. Ent. Soc., 1893, p. 55.
† The Entomologist, 1896, p. 240.

The reason of this is probably that a temperature of 0.5° is so low that it paralyses all tissue changes in the pupæ, and at the end of ten weeks the stage of development is no further advanced than at the beginning. Thus the time of emergence of these pupæ, after transference to a temperature of 32°, was just as long as for those kept only at this temperature.

Arguing from his experiments on two moths, *Selenia illustraria* and *Ennomos autumnaria*, Merrifield * came to the conclusion that, in their case at least, the *markings* were chiefly affected by the temperature experienced during the earlier part of the pupal period, whilst the *colouring* was "chiefly affected during the penultimate pupal stage, *i. e.*, before the colouring of the imago begins to show." A low temperature during this latter period causes darkness, and a high temperature the opposite effect. Thus, by difference of treatment, it was found possible to obtain from the same brood individuals showing (1) summer markings with summer colouring; (2) summer markings with an approach toward spring colouring; (3) spring markings with summer colouring, and (4) spring markings with almost spring colouring.

We see, then, that in some cases seasonal dimorphism is a direct response to temperature, or is a somatic modification, whilst in other and perhaps the majority of cases it is only indirect, the temperature acting as a stimulus to arouse a blastogenic variation. When the response is direct, low temperature generally induces a darkening of colour, as, for instance, in *V. urticæ*, *Lasciocampa quercus* (and *callunæ*), *Arctia caja* and *E.*

* Trans. Ent. Soc. 1891, p. 55.

autumnaria. In these forms, the darkening is caused either by the general colour being obscured, or by the size and general intensity of the dark markings being increased, or by both conditions.* In *P. phlæas*, as we have seen, low temperature causes a lightening of colour. When the response to temperature is indirect, the effect is as often as not in one direction as in the other, and there are generally more considerable changes in the markings, as well as in the general colouring.

Upon the higher animals temperature probably acts but seldom as a direct cause of variation. The white coat which many quadrupeds develop on the approach of winter in northern and arctic climates is probably in great part a seasonally adaptive change, but it may also be to a certain extent the immediate, though perhaps only indirect, response to cold. This seems to be proved by an observation of Sir J. Ross on a Hudson's Bay Lemming.† This animal was protected from the low temperature by keeping it in the cabin, and had in consequence retained its summer coat through the winter. On exposing it in a cage on deck, where the temperature was 30° below zero, the fur on the cheeks and a patch on each shoulder became perfectly white during the first night. After another day's exposure "the patches on each shoulder had extended considerably, and the posterior part of the body and the flanks had turned a dirty white. . . At the end of a week it was entirely white, with the exception of a dark band across the shoulders, prolonged posteriorly down the middle of

* *Vide* Merrifield. Trans. Ent. Soc. 1892, p. 33.

† Appendix to Second Voyage. Nat. Hist., p. xiv., 1835. Quoted from Poulton's "Colours of Animals," ed. i. p. 94.

the back." No further change took place, and the animal died of the cold a few days later. Examination of the fur showed that only the tips of the hairs had become white, so that on cutting these off, the coat regained its original dark colour.

The observations of F. H. Welch * on the American Hare (*Lepus Americanus*) throw further light on the nature of the change. Early in October the whiskers and a few of the longer hairs on the back were observed to become white at the tip or throughout. During November a new and rapid growth of stiff white hairs appeared on the sides and back, these hairs being easily distinguishable from the autumnal hairs which were gradually turning more and more white, in that they were invariably white throughout. We have in this animal, therefore, a new white crop of hairs of gradual growth, or a blastogenic variation, stimulated to develop under stress of cold, and a rapid and direct transmutation of parts of the dark hairs to white; *i. e.*, a somatic modification. Professor Poulton † explains this latter change as an indirect influence of cold upon the nervous system which presides over the nutritive and chemical changes involved in the growth of the hair. This probably leads to the production of large numbers of gas bubbles in the hairs, and thereby induces an apparent whiteness, in spite of the fact that the pigment is still present. In that the tips of the hairs are first affected, however, rather than the bases, it seems to me possible that the cold acts directly on the hairs themselves, and not indirectly through the nervous system.

* Proc. Zoöl. Soc. 1869, p. 228.
† "Colours of Animals," p. 100.

It should be pointed out that some animals, such as the sable, musk-sheep, and raven, retain their dark colour throughout the Arctic winter, so that the reaction of the above-mentioned animals to cold, whether direct or indirect, is a special and not a general phenomenon.

Light. The effect of light upon growth, especially in plants, is well known to be very considerable. One might infer, therefore, that differences in the intensity of the light to which an organism is subjected would form a potent cause of variation. Such is actually the case among members of the Vegetable Kingdom, though only exceptionally so among those of the Animal Kingdom.

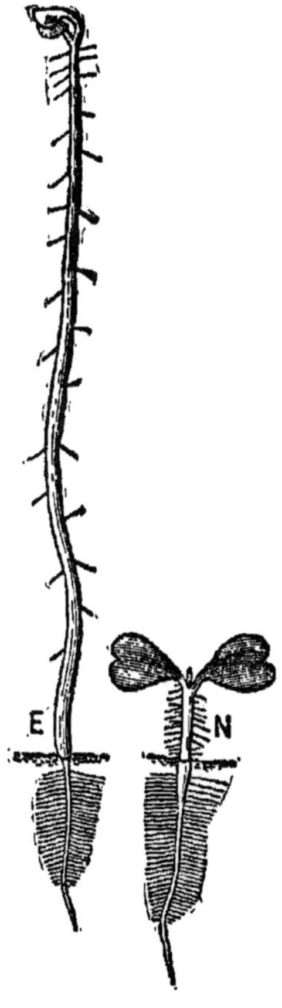

FIG. 23.—Seedlings of *Sinapis alba*. *E*, reared in the dark. *N*, reared in ordinary daylight.

If plants be allowed to grow in absolute darkness, they, as a rule, become very much elongated in form, whilst their leaves are small and ill-shaped. The accompanying figure shows the relative growth of two seedlings of *Sinapis alba* of the same age, one of them reared in the dark; and the other in ordinary daylight.* Sachs found that potato tubers grown in darkness for 53 days produced sprouts from 150 to 200 mm. high, whilst similar ones grown in day-

* From Strasburger, Noll, Schenck, and Schimper's "Textbook of Botany." Quoted from Davenport's "Experimental Morphology," p. 418.

light were only 10 to 13 mm. high. Again, he found that the hypocotyl of the buckwheat (*Fagopyrum*) reached a height of 35 to 40 cm. in the dark, whilst it grew only to 2 or 3 cm. when freely exposed to light. K. Goebel* has shown that if cactuses are cultivated in darkness, their form changes completely. The young shoots are rounded, and fail to show the angular irregularities of form which increase the surface capable of effecting assimilation under the influence of light.

Darkness conduces to increased growth, therefore, or conversely, light tends to retard growth. That this is the case is well shown by an observation of Wiesner.† This observer exposed seedlings of the vetch (*Vicia sativa*) under a glass globe to sunlight for 7½ hours. When placed horizontally, so as to get the full force of the sun's rays, no growth whatever occurred, but when placed vertically, so that the growing part of the seedling was more or less protected by its leaves, there was an increase in height of about .8 mm. On the other hand, a control seedling kept in a darkened globe grew about 2.8 mm. in the same period. This retarding effect of light is not universal, however. It is practically absent in some cases, as of the yam and of a wild gourd (*Bryonia*), and in those plants whose rapidly growing parts are sheltered from the sun's rays by protecting coverings it is but little evident. Still Sachs' conclusion as to the effect of daylight on growth probably applies with greater or less force to the majority of plants. Thus he found‡ that during the night the growth

* Flora, lxxxi. p. 96.

† Davenport's "Experimental Morphology," p. 41.

‡ Arb. aus der Bot., Inst. Würzburg, i. p. 99.

gradually increases, and reaches a maximum at daybreak. It then diminishes to a minimum a little before sunset, after which it rises again.

It is not to be imagined that because daylight retards growth it is unfavourable to the proper development of a plant. For instance, Karsten * found that whilst a kidney bean reared in the dark for a month or two weighed 20 per cent. more than one reared in the light, yet the leaves did not weigh a fifth as much. Again, Clayton † allowed six bean plants to grow in a spot where they would catch all the sunshine of the day, whilst six other similar plants were protected by a boarding, which effectually screened off the sun. When freshly gathered in October, the weight of the beans and pods of the exposed plants was to that of the protected as 99 : 29, whilst the weight of the dry beans was as 16 : 5. The next year, the weight of the fresh beans and pods obtained from the sunshine-grown seed of the previous year was half as much again as in the case of the plants from shade-grown seeds, in spite of the fact that all of the plants were now grown in sunshine and under precisely similar conditions. "In the fourth year plants with an exclusively shady ancestry produced flowers, but failed to mature fruit."

The intensity of the light to which a plant is exposed may considerably affect its form and structure, as well as its rate of growth. Thus the effect of direct sunlight, as compared with diffused light, on the absolute

* Landw. Versuchs-Stat., xiii. p. 176. Quoted from Davenport's "Experimental Morphology," p. 419.

† Nat. Science, xi. p. 12.

size of leaves has been shown by Stahl * to consist chiefly in a reduction of the leaf surface. Accompanying this there is usually an increase in the thickness of the leaf. In addition to the reduction of size, Scott Elliott † has shown that there may be a considerable change in the form of the leaves, owing to the reduction in the length of the exposed leaves being much greater than the reduction in the breadth. The accompanying table shows the average ratio of length to breadth in from 50 to 100 leaves of various grasses and plants, which were collected in the one case from sheltered and shady places, and in the other from the most exposed and driest spots known:

NAME OF SPECIES.	SHELTERED SPECIMENS.	EXPOSED SPECIMENS.	PER CENT. REDUCTION.
Stenotaphrum glabrum,	15.8	6.1	61.4
Paspalum distichum,	23.0	11.0	52.2
Cynadon dactylon,	7.3	4.5	38.4
Eragrostis ciliaris,	32.4	17.5	46.0
Cenchrus echinatus,	22.5	19.8	12.0
Microrhyncus sarmentosus,	6.8	4.7	30.9
Lobelia Scævola,	2.2	1.6	27.3
Psiadia dodonæfolia,	10.2	11.4	+11.2
Helichrysum emirnese,	9.4	5.6	40.4
Spermacoce globosa,	6.0	4.2	30.0
Lycium capense,	1.9	1.9	0.0
Brexia madagascariensis,	1.8	1.5	16.7
Camptocarpus, sp.,	4.1	3.6	12.2
Periploca ovata,	1.6	1.5	6.2
Commelina nodiflora,	3.2	2.9	9.4
Tanghinia venenifera,	4.7	3.6	23.4
Brachystephanus cuspidatus,	1.7	1.7	0.0
Monimia, sp.,	2.0	1.8	10.0
Sida carpinifolia,	2.6	2.2	15.4
Vinca rosea,	2.3	2.3	0.0

* Jenaisch. Zeit., Bd. xvi. p. 102, 1882, and Bot. Zeit., 1880.
† Proc. Linn. Soc., vol. xxviii. (Botany), p. 375, 1891.

Of these twenty different species of grasses and plants (collected in Madagascar), we see that the exposed specimens had a reduced leaf length ratio in 16 cases, whilst in only one was the length actually increased. The average reduction for the whole series amounts to 21.0 per cent., or is very considerable. Similar results to these have also been obtained by Sorauer.*

Upon members of the Animal Kingdom the direct effect of light is not nearly so considerable. Yung † found that tadpoles exposed to daylight during the first 25 to 60 days of development were about 16 per cent. larger than those kept in absolute darkness. He found also that eggs of the sea-trout, if reared in the light, hatched a day earlier than if reared in the dark, whilst pond snails (*Lymnæa stagnalis*) hatched in 27 days in the light, as against 33 days when in the dark.‡ It is possible, however, that these effects were due rather to the presence or absence of heat rays than those of light.

The most important influence of light in the production of variations in animals lies in its connection with the phenomena of pigmentation. Absence of light leads to diminution or even total abolition of pigmentation, whilst its presence leads to an increase in some degree proportionate to the intensity of the light. This, at least, is the more or less direct action of light. The indirect action, through the intermediation of the nervous system, is, as a rule, exactly the reverse. A

* Wollny's "Forschungen a. d. Geb. Agricultur," Bd. ix.
† Arch. Zoöl. Expér. et Gén., vii. p. 251.
‡ Davenport's "Experimental Morphology," p. 426.

well-known instance of the direct action of light is found in the bronzing of the human skin following on undue exposure to the sun; but to what extent are we entitled to refer the black skin of inhabitants of the tropics to a similar, but more pronounced, action? Eimer * is of the opinion that the effect is the direct result of the more intense light and heat. Thus he found that in passing down the Nile valley from the Delta to the Soudan, the natives gradually became more and more dark-skinned, the further south they lived. The increased light and warmth, according to Eimer, lead to a greater flow of blood to the skin, and the consequent deposition of pigment. This effect is inherited, and has become a constant character. There is, of course, no warrant for laying down the law with such assurance as this, for one can easily imagine several other equally possible and plausible explanations to account for the facts. For instance, pigmentation may be correlated with a greater resistance to the climate of hot countries, or with greater physical strength, and may have been increased by sexual selection. Still, Eimer's explanation may contain a distinct modicum of truth, and I hope to prove in a subsequent chapter that the heritableness of acquired characters such as increased or decreased pigmentation may be deduced without assuming anything further than the present state of knowledge legitimately warrants us in doing.

The diminution or disappearance of pigmentation following upon withdrawal of light is best illustrated by reference to the well-known cave animals. Of these,

* "Organic Evolution," p. 87.

one of the most interesting is *Proteus anguineus*, which is found in the subterranean caves of the Karst Mountains about Adelsberg. This amphibian is almost white, but if kept for some time in the light, it gradually becomes pigmented. Pigment cells are, in fact, still present in its skin,* and in all probability these are directly stimulated to exert their function by the action of the light.

A similar effect of exposure to light has been demonstrated by Cunningham † for the under surface of the flounder (*Pleuronectes flesus*). This surface is normally quite white, but by keeping young flounders for nearly four months in a glass dish illuminated from beneath by a mirror placed at a proper angle, Cunningham found that 10 out of the 13 specimens experimented with developed black and yellow chromatophores. Three of the specimens showed well-developed bands of pigment, similar to those of the upper side, over the area occupied by the muscles of the longitudinal fins. Subsequently, Cunningham and MacMunn ‡ succeeded in keeping flounders alive under these conditions of illumination for from 9½ months to nearly two years. They found that the amount of pigment steadily increased with the duration of the exposure, so that ultimately almost the whole of the lower side might become pigmented. This colouration was (histologically) of exactly the same kind as that of the upper side in normal specimens, though it was never by any means so marked. Its production is more re-

* *Vide* Poulton's "Colours of Animals," p. 91.

† Zoöl. Anzeiger, xiv. p. 27, 1891.

‡ Phil. Trans. 1893, B. p. 765.

markable than in the case of *Proteus*, in that pigment cells are entirely absent from the skin of the lower side of the normal Flounder.*

The observations of List † upon certain Lamellibranch Molluscs afford evidence as to the effects both of decrease and increase of illumination. List noticed that various species of *Mytilus* (*gallo-provincialis* and *minimus*), which had been collected in caves, were distinctly less pigmented than usual. In fact, those obtained from the extreme ends of the two grottoes underneath the ruined Palace of Donn'Anna in the Bay of Naples were all of them pale or colourless. Individuals of the same species were also found in the dark underground tanks of the Zoölogical Station, and here again they were little, if at all, pigmented. Specimens of *M. minimus* were even found in the pipe through which the water is pumped from the sea into the Aquarium, and these were characterised by an absolute want of pigmentation. The converse observations were made upon *Lithodomus dactylus*. These molluscs, which are usually concealed in borings in the sand of the sea bottom half a metre deep, are pigmented only at the tip of the foot and the edge of the anal siphon, these being the only parts at all exposed to light. After keeping specimens for a year in a glass vessel exposed to daylight, however, the whole surface of the anal siphon became coloured an intense red brown, whilst the imperfect branchial siphon, the border of the mantle, the whole of the foot, and the other exposed parts, were pigmented also.

* *Ibid.*, p. 767.

† Arch. f. Entwickelungsmechanik, Bd. viii. p. 618, 1899.

Again, Viré * has obtained somewhat similar results in his observations on the Fauna of subterranean caves and streamlets in France. As regards the Crustacea, he found that *Niphargus virei*, which is of a rose colour, after a few weeks' exposure to light becomes covered with brown spots, and thus undergoes a rapid return to its ancestral form. On the other hand *Gammarus puteanus*, when kept for ten months in the tanks of an underground laboratory, began to lose its gray-green pigment, and after twenty months most specimens had entirely lost it. Again, the common *Gammarus fluviatilis*, when kept for fifteen months underground, developed organs of touch and smell which attained nearly half the size of those exhibited by the true cave *Niphargus*.

The observations which have been made on *Amblyopsis* (a fish found in the caves of the Mississippi Valley) do not agree with the above results. Thus Eigenmann † states that the pigment is very abundant when the young fish are two months old, but even when these fish are kept in light during growth, they show a decrease and not any increase of pigmentation, so that a ten-months' fish was found to have taken on the exact pigmentless condition of the adult. Both the pigmented condition and the subsequent depigmentation are hereditarily transmitted, therefore, and seem practically unaffected by changes of environment acting through a single generation.

* "La Faune souterraine de France," Paris, 1900; *vide* Abstract by P. Kropotkin in Nineteenth Century, September, 1901, from which this reference is taken.

† Biological Lectures, Wood's Holl, 1899, p. 124.

The direct dependence of pigmentation on light seems to be proved by the generality of the reverse phenomenon as observed in cave animals. Whenever light is totally excluded, the pigmentation appears to become diminished or abolished, whatever class of the Animal Kingdom the individuals belong to. Thus there have been found more or less unpigmented Cœlentera, Worms, Crustacea, Myriapoda, Arachnida, Coleoptera, Fish, and other animals in the various subterranean caves of Europe and North America.* However, the abyssal fauna of the ocean, occurring at depths such that (presumably) no light can penetrate, includes numerous species which are just as much pigmented as those exposed to light. Thus Faxon † divides deep sea Crustacea into two types; (1) those living in the bottom mud, which are mostly pale in colour, and often blind; (2) those which swim freely, have well-developed eyes, and are coloured bright red. He considers that this red colour is due to the absence of light at these profound depths, for S. Jourdain ‡ has shown that two different species of Crustacea, which are brown when exposed to light or partial darkness, become red when placed in total darkness. MacCulloch and Coldstream have suggested a "theory of Abyssal Light" to account for the existence at profound depths of these Crustacea, and of the Fish, Mollusca, Crabs, and other animals with well-developed eyes. This hypothesis "consists essentially in the idea that light diffused by

* Packard, Memoirs of National Academy of Sciences, iv. p. 3, 1888.

† Mem. Mus. Harvard, xviii. p. 251.

‡ Comptes Rendus, lxxxvii. p. 302, 1878.

phosphorescent creatures is capable of taking the place of sunlight in those depths which the rays of the sun cannot penetrate." *

The more striking and considerable effects produced by light in members of the Animal Kingdom are mostly confined to cases of so-called "Variable Protective and Aggressive Resemblance," or reaction to the colour of the surroundings which either protects the animals from their enemies, or assists them to capture their prey.† Such a reaction is rarely, if ever, a direct response to light of the superficial tissue cells as a whole, or even of the sensitive pigment cells in the skin which have been gradually formed in the course of evolution through the agency of Natural Selection and other processes. It is an indirect effect produced by the intermediation of the nervous system. This was first proved to be the case by Brücke ‡ for the chameleon, and by von Wittich § for the frog. The latter observer regarded the variations in colour as probably reflex in their nature, he attributing them to a peripheral ganglionic apparatus in the skin itself. A few years later Lord Lister ‖ took up the problem and correctly solved it, he concluding that in *Rana temporaria* "the eyes are the only channels through which the rays of light gain access to the nervous system so as to induce changes of colour in the skin." The very conspicuous changes which can be produced in this manner may be illus-

* Quoted from Semper's "Animal Life," p. 85.

† *Vide* Poulton's "Colours of Animals," pp. 81 to 158.

‡ IV. Bd, d. mathemat. naturwiss., Classe d. Kaiserl. Acad. d. Wissenschaft, Wien, 1852.

§ Muller's Archiv, 1854, p. 41.

‖ Phil. Trans. 1858, p. 627.

trated by another quotation from Lord Lister's paper: "A frog caught in a recess in a black rock was itself almost black; but after it had been kept for about an hour on white flagstones in the sun, was found to be dusky yellow, with dark spots here and there. It was then placed in the hollow of the rock, and in a quarter of an hour had resumed its former darkness. These effects are independent of changes of temperature." These changes of colour have been shown by Brücke, von Wittich, Lister, and others to be due to the pigment granules of certain stellate cells in the skin varying in their degree of concentration towards the centre of the cell, and in their diffusion peripherally through the branching processes. These pigment cells are often of different colours and are arranged in layers, so that widely different effects may be produced by varying degrees of concentration in them.

That the reflex mechanism takes its origin in the eye, which is stimulated by the light reflected from the animals' surroundings, was proved by Lord Lister in the following manner: He found that a frog with its eyes removed was totally unaffected by the colour of its surroundings. The nervous system still retained the capacity for acting on the pigment cells, however, as the frog, originally dark, became extremely pale after struggling violently to escape. It was then placed in a bright light, but within half an hour became almost coal black again. Occasionally protectively coloured animals are found in nature showing a total want of adjustment to the colour of their surroundings. Thus Pouchet* noticed that one

* Quoted by Semper, "Animal Life," p. 95.

single plaice out of a large number upon a bright sandy surface was dark-coloured, and Nicoll * noticed that in addition to the light-coloured trout usually seen in a chalk stream in Hampshire, very dark individuals occasionally appeared. In both instances, however, it was proved that the fish were blind, and therefore unable to respond to the stimulus of reflected light.

Besides the amphibia, fish, and reptile mentioned, many other animals belonging to the same groups exhibit a similar power of rapidly adapting their colour to that of their surroundings. The power is also possessed by many invertebrate animals. It is probably very common among Crustacea, and some cuttle fish can modify their colours with extreme rapidity. In *Octopus vulgaris* the protective resemblance is very striking, and so completely is it under the control of the nervous system that I have seen an individual change its colour from a dirty white to a dark brown in less than a second. It is amongst the Lepidoptera, however, that our knowledge has been furthest advanced. The power of adaption has so far been proved to exist in this group during the larval and pupal stages only, though it is probable that a relatively small number of perfect insects also possess it.† Again, it is present only in such pupæ as are exposed, and has been found wanting in those of moths which are as a rule either buried in the earth or concealed in opaque cocoons. Professor Poulton ‡ has shown, however, that the pupa of the Swallow-tailed moth forms an ex-

* "Colours of Animals," p. 86.
† "Colours of Animals," p. 110.
‡ *Ibid.*, p. 111.

ception to this rule. Also he has found that the cocoons themselves may undergo protective colouration.

The first recorded instance of variable protective resemblance in Lepidoptera is due to T. W. Wood,* who in 1867 demonstrated it for the pupæ of the Large and Small White butterflies (*Pieris brassicæ* and *P. rapæ*). A few experiments were made from time to time by other observers, but it was not until 1886 that they were undertaken systematically on a large scale. This was done by Professor Poulton, who obtained most striking results.† Working upon over 700 chrysalides of *Vanessa urticæ* (Small Tortoiseshell), he found that pupæ placed against black surroundings became as a rule extremely dark, whilst against white surroundings "not only was the black colouring matter as a rule absent, so that the pupæ were light-coloured, but there was often an immense development of the golden spots, so that in many cases the whole surface of the pupæ glittered with an apparent metallic lustre." Against a gilt background a much higher percentage of gilded chrysalides was obtained, and this led Professor Poulton to suggest that in its original habitat the larvæ pupated either against glittering micaceous rocks which had a somewhat metallic appearance, or against dark rough weathered rocks, and that they had acquired the power of protectively resembling either of these surfaces. In that such metallic looking rocks occur over a comparatively limited area, whilst the species has a consider-

* Proc. Ent. Soc., vol. xiii., 3d series, p. xcix. 1867.

† Phil. Trans. 1887, B. p. 311, *also* "Colours of Animals," p. 119 *et seq.*

able range, Dr. A. R. Wallace * considers Professor Poulton's suggestion rather improbable; still it should be noted that the Small Tortoiseshell almost invariably seeks mineral surroundings for the pupal period, and very rarely becomes a chrysalis on its food plant.

The time at which the colours are determined was found to be especially during the resting stage of the caterpillar, just before pupation, and to a less degree during the onset of the pupal stage, when the caterpillar hangs head downwards, suspended by its last pair of claspers. The former stage lasts about 15 hours, and the latter about 18, and at the end of it the skin splits along the back of the head, and the chrysalis becomes exposed. The reaction of the skin of the larva to the colour of its surroundings was proved by some ingenious experiments to be an indirect one, effected probably through the medium of the nervous system. Thus when a larva, during the onset of the pupal stage, was so placed that part of it was illuminated by a gilded background, and part by a black one, parti-coloured pupæ were never obtained. The effective results were produced by that colour to which the larger area of skin had been exposed.

We see, then, that in the development of certain Lepidoptera there is a period, lasting only a day or two, during which an extreme sensitiveness to the colour of the surroundings is present, and we have also seen that during the pupal period there may be a great sensitiveness to the temperature of the surroundings. These cases therefore form exceptions to the conclusion arrived at in the last chapter, viz., that reaction to environment

* "Darwinism," p. 198.

diminishes regularly with progress of development. It is obvious, however, that both these capacities for reaction are quite unusual, and have been specially acquired for a special purpose. In all probability the organisms are not more sensitive to environmental conditions in general at these periods than they are at the earlier ones; in fact, they are probably very much less so, in that the growth has almost ceased.

In certain caterpillars the existence of a variable protective resemblance has long been recognised, several instances of the phenomenon being collected by Meldola * in 1873. For example, the larvæ of the genera *Smerinthus* and *Sphinx*, which are green when feeding on their respective food plants, become brown previous to pupation, when the caterpillars are crawling over the ground to find a suitable burying place. Again, the Geometer *Acidalia degeneraria* is greenish brown in the summer, but changes to a rusty brown in the autumn, at the period preparatory to hibernation. Some years later, Meldola recorded an observation by Mr. E. Boscher,† relative to the larvæ of *Smerinthus ocellatus* (Eyed Hawk Moth). These larvæ were noticed to be of a whitish green colour when feeding on one species of willow, and of a bright yellowish green when feeding on another species, these colours being, on the whole, protective. It was generally believed that such variability in the colour of caterpillars is due to the direct chemical effect of the different kinds of leaves eaten, but Professor Poulton,‡ by his experiments on the

* Proc. Zoöl. Soc. 1873, p. 153.
† Weismann's "Studies in the Theory of Descent," 1882, p. 241.
‡ "Colours of Animals," p. 149.

larvæ of *Smerinthus*, proved that it was the *colour* of the leaves, and not their food quality, which provoked the change. Thus he sewed leaves together, "so that the caterpillars were exposed to the colour of the upper or of the under side alone, although they ate the same leaf in both cases. In other instances the bloom was rubbed off the under sides of some leaves, whilst others were left normal."

More striking cases of protective resemblance were obtained by Professor Poulton for various *Geometræ* larvæ. Larvæ surrounded by the leaves on which they fed, became, in the majority of species, light brown or light gray in colour. If, however, an abundance of twigs had been mixed with the leaves of the food plant, they became dark in colour. The larvæ of the Peppered moth (*Amphidasys betularia*) afforded the most striking result of all, for when reared amongst green leaves and shoots they became bright green without exception, whilst in the presence of dark brown twigs they nearly all assumed a corresponding colour.

The influence of the surroundings acts only very slowly upon the colour of the caterpillars, the coloured part being "actually built up of the appropriate tint." Probably this is the result of light stimuli acting on the surface of the skin, and not reflexly through the eye. Thus painting the eyes (ocelli) with opaque varnish led to no diminution of reaction.

CHAPTER VIII.

THE EFFECT OF MOISTURE AND OF SALINITY.

Effect of humidity of soil on plant growth—Effect of dry and moist surroundings on characters of plants—Desert plants and Aquatic plants—Effect of moisture on Lepidoptera and on Molluscs—Characters of maritime plants probably due to saline environment—Conversion of *A. salina* into *A. milhausenii* and into *Branchipus*—Effect of increased salinity on characters of the cockle—Influence of salinity on rate of growth of Tubularians, and on size of sea-urchin larvæ.

In that the presence of water is absolutely necessary to enable living organisms to exhibit activity, and very probably, indeed, to enable them to retain vitality at all—for even spores contain a small percentage of water—so we should conclude that differences in the amount of water in the environment of the organisms would form a fertile source of variation. Such is, in fact, the case in the Vegetable Kingdom. For example, the amount of water in the soil has a considerable influence on the rate of growth, as is shown in the table given below. These figures, which were obtained by Hellriegel,* represent the amount of dry substance contained in the grain and chaff of barley which had been reared in soils containing various percentages of the saturation quantity of moisture. We see that the rate of growth varies but little until the humidity falls below 30 per cent., and then it diminishes so rapidly that with a hu-

* Quoted from Davenport's "Experimental Morphology," p. 353.

midity of 10 per cent. it has almost ceased. In the observations made by Gain,* the fresh weight of the entire plant was determined. Seeds of various species were planted in soil containing either from 3 to 6 per cent. of water, or from 12 to 16 per cent. Growth was more rapid in the damp than in the dry soil, so that the weight of the full-grown plant was 1.12 times greater in the radish, 2.23 times in the bean, and 2.7 times in the flax.

HUMIDITY OF SOIL.	PRODUCTION IN DRY SUBSTANCE.	
	GRAIN.	CHAFF.
per cent.		
80	8.8	9.5
60	10.0	11.0
40	10.5	9.6
30	9.7	8.2
20	7.7	5.5
10	.7	1.8
5	.0	.1

The effect of a dry soil and atmosphere is well shown by the characters of desert plants. These are stunted in growth, and are of a nearly uniform gray colour, owing to their intense hairiness. The leaves are more fleshy, and there is a great tendency to the formation of spines. That these characters are in part at least the direct result of want of water is shown by the fact that they may disappear if an abundance of water is supplied. Thus *Ononis spinosa.* L., if grown in a rich, well-watered soil, or in a moist atmosphere, gradually loses its spines, those first formed under the new

* Ann. Sci. Nat. Bot. (7), xx. p. 63, 1895.

conditions being much reduced in size and rigidity.* Lothelier † has made numerous observations in which individuals of the same species were placed side by side, some exposed freely to the air, and others kept moist under a glass shade by a vessel of water. He found that, for instance, *Berberis vulgaris* bore non-spinescent leaves in a moist atmosphere, but spines and spines alone in a perfectly dry one. Again, the shoots which in *Lycium barbarum*, *Ulex europæus*, etc., would normally have formed thorns by arrest in development and sclerosis, in a very damp atmosphere continued to grow, and elongated into leafy branches. Microscopical examination showed that in the moist atmosphere the parenchyma was only imperfectly differentiated into spongiform and palisade tissue, whilst in dry air there was a great arrest in the area of parenchymatous tissue, but the palisade cells were well developed, and there was a special consolidation of fibrous tissues. Again, the common water-reed, *Phragmites communis*, when growing in the unirrigated areas of the Nile Valley, forms a stunted growth, with very short and sharp-pointed leaves. "Close to the Nile, however, in Rhoda Island, it grows nine or ten feet high, with long leaves almost exactly like the plants in English rivers." ‡ The effect of drought upon *Dioscorea batatus* (Yam) has been carefully studied by Duchartre.§ Though not allowed to have any water, some tubers of this plant

* Rev. G. Henslow, "The Origin of Plant Structures," p. 40.

† See also "Recherches anatomiques sur les épines et les aiguillons des plantes," Lille, 1893.

‡ "Origin of Plant Structures," p. 41.

§ Bull. de la Soc. Bot. de France, 1885, p. 156.

produced long shoots. The stem was more slender than usual, but excessively rigid, owing to the reduction of the parenchymatous tissues, and the predominance of the elements of consolidation. The leaves were small and undifferentiated, and the stomata undeveloped.

Many other instances showing the relations between floral structures and arid surroundings have been collected by Henslow in his book on the "Origin of Plant Structures," where the subject is dealt with *in extenso.*

The effects of a very great increase in the humidity of the surroundings, such as is experienced by plants which actually live in water, lie, as might be expected, in a very different direction. That the peculiar characters of aquatic plants are in considerable measure the direct effects of their peculiar environment, is proved by the fact that plants, normally terrestrial, often develop such characters when grown in water. Thus Costantin found that under such conditions a diminution in the number of the vessels of the fibro-vascular system of the stem invariably occurred. For instance, in *Vicia sativa* (Vetch) the middle of the stem of the aquatic form of the plant had only 38 vessels, whilst the aërial form had 47. In *Ricinus communis* (Castor-oil plant) the aquatic form had 10, as against 21; in *Faba vulgaris* it had 2 at the sides and 15 at the angles, as against respectively 5 and 36. The pseudo-aquatic forms thus tended towards true aquatic plants, in which the fibro-vascular system is always more or less degenerate. Again, Costantin found that there was an increase in the lacunæ, when stems normally aërial are kept submerged, just as the aquatic form of amphibious plants is found to have more of such lacunæ than the aërial. As regards the leaves, it is well known that when aërial and floating leaves are present on the same aquatic plant, they differ greatly in structure, and as a rule also in form, from the submerged leaves. In *Ranunculus heterophyllus* and *Cabomba aquatica*, for instance, the floating leaves are more or less rounded, whilst the submerged ones have dissected and filiform segments. In *Hippuris* (Mare's tail) the aërial and floating leaves are short, and in *Callitriche* rounded, but the submerged leaves of both are linear or ribbon-like. In all cases the submerged leaves are of a more delicate texture, more or less translucent, and of a brighter green colour than the others. They show degradation of anatomical structure in every part, the cuticle and stomata disappearing, whilst the chlorophyll grains and the mesophyll are greatly reduced in quantity.* Even better evidence of the direct relation between environment and character is afforded by certain other observations of Costantin. Thus he found that he could change the form of *Hippuris* at will, "by transplanting an aquatic plant on to land, and *vice versa;* all the leaves produced under water were long, undulated, and delicate; whereas those in air were short, erect, and firm." Again, he found that the leaves of *Sagittaria* (Arrowhead), when deeply submerged, are soft and flexible, and may reach a length of over six feet, but when developed in air they are short and erect.

When a leaf is full grown, sudden change of environment kills it, aërial leaves perishing under water, and

* *Vide* Henslow's "Origin of Plant Structures," chap. viii., from which the greater part of this paragraph is drawn.

aquatic ones perishing in air; but if it is only in the course of its development, it can adapt itself to a changed environment. Thus, if a half-formed floating leaf of *Ranunculus heterophyllus* or of *Sagittaria* is submerged, "it is at once arrested, and begins to re-adapt itself to water." There are some species, however, such as many algæ, which show no power of adaptation, and can only live entirely under water.

The converse experiments of growing aquatic plants on land afforded equally striking results. For instance, it was pointed out by Godron,* as long ago as 1839, that whilst *Ranunculus aquatilis* (Water crowfoot), when wholly submerged, has all its leaves delicately laciniated, yet "if the plant is able to send some of its leaves to the surface, they float and assume a very different form, being kidney-shaped and lobed. The same plant when growing entirely out of water presents a very different appearance; the stem is short, much divided into branches, which bear a large number of small leaves, cylindrical, much divided, and somewhat thick. If it were not for the floral organs, one would certainly believe in two or three species." Again, Costantin grew a plant of the aquatic form of *Peplis Portula* on land, and found that the internodes were changed from their elongated form to a short one. The septa of the cortical parenchyma of the stem remained homogeneous, instead of being hollowed out into secondary lacunæ, and also the number of vessels was increased. Thus there were 53 vessels in the land form of *Peplis Portula*, instead of 25; 12 instead of 4 in *Callitriche*, and 57 instead of 18 in *Nasturtium*.

* Quoted from De Varigny's "Experimental Evolution," p. 97.

It may have been noticed that in speaking of these adaptations of terrestrial plants to water, and of aquatic plants to land, it has been more or less tacitly assumed that the effects observed were due to the direct influence of the surroundings on the tissues. It is of course possible that they are partly or even largely indirect, and that the change of habitat merely calls up latent characters long since possessed by the ancestral plants which lived in similar surroundings.

Upon members of the Animal Kingdom, observations as to the effect of moisture are exceedingly meagre. This is probably attributable to the fact that in most cases a direct effect is either slight or wanting. Thus Merrifield * could not observe any influence upon the pupæ of certain Lepidoptera (E. *autumnaria* and *S. illustraria*), nor could Standfuss upon those of certain other species. Koch,† however, came to the conclusion that a long period of dry or moist weather might exercise a considerable influence on the size of the succeeding generation. Immediately after a continuously dry summer, butterflies are always smaller than after a moist one. Likewise also the second generation of *Argynnis selene*, which takes flight in the height of summer, is always smaller than the spring generation; but it seems to me highly probable that these effects are of an indirect nature, dependent, perhaps, on changes effected by the moisture in the vegetation on which the larvæ feed.

Leydig ‡ has endeavoured to trace a connection be-

* Trans. Ent. Soc. 1891, p. 163.

† Quoted from Eimer's "Organic Evolution," p. 152.

‡ "Organic Evolution," p. 97.

tween the moisture in the environment and the darkness of colouring of certain animals. Thus he observed that Molluscs such as *Arion empiricorum* (common slug), *Helix arbustorum*, *Succinea Pfeifferi*, and *Helix circinata* became darker than usual in moist localities. He observed a similar condition also in certain Amphibia and in *Lacerta vivipara*. However, Eimer observed just the reverse condition in *Arion*, finding it darker upon the heights, where there was little water, than in well-watered valleys. In any case, the effect is probably an indirect one, acting through the vegetation.

The Effect of Salinity. The effect of salinity upon members of the Vegetable Kingdom is well illustrated by the peculiarities of form and structure possessed by maritime plants. That these characters are at least in part the direct effect of the salinity of the soil and atmosphere, is proved by comparison of plants growing near the sea-shore with individuals of the same species growing inland. Thus Lesage * has investigated no less than 85 different species. He found that in 54 of them the leaves were thicker in the maritime individuals than in the inland ones, they being about four times as thick in *Cakile maritima* and *Silene maritima;* in 27 there was no apparent difference, and in 4 they were thicker in the inland individuals. With regard to the mesophyll, there was no noticeable change in 11 species, but in all the other shore plants the palisade cells were more numerous or attained greater thickness, and at the same time the interspaces underlying the stratum of palisade cells were much reduced.

* "Influence du Bord de la Mer sur la Structure des Feuilles," Rennes, Oberthur, 1890. *Also* Rev. Gén. de Bot., tom. ii. p. 54.

Changes in the epidermis were much less frequent, there being no appreciable difference in 31 plants. In 23 of the shore plants the cells were larger, however, they being three to five times as large in *Beta vulgaris* and *Silene maritima*. In four instances these cells were larger in the inland plants. With regard to the chlorophyll, there was no difference in some cases, but in others it was marked. Thus in *Thesium humifusum* and *Cakile maritima* the grains were much smaller in the maritime plants, and in other species the number of grains was reduced.

Even more conclusive evidence of the direct effect of salinity in producing these peculiarities of structure has been afforded by experiment. Lesage cultivated various plants under similar conditions except that some of them were watered with water containing common salt, and he found that characters were developed similar to those exhibited by maritime plants. In *Pisum sativum* the leaves increased in thickness, palisade cells became larger and more numerous, whilst the intercellular spaces and the chlorophyll diminished. *Lepidium sativum* (Garden Cress) gave even more marked results. The palisade tissue was more developed and possessed an extra layer; the lacunæ were less pronounced, and the chlorophyll less abundant. On sowing the seeds of this plant a second year, moreover, and again treating the plants with salt water, a still more marked result was obtained, it appearing as if the alteration in the tissues of the second generation was carried on more or less from the point gained in the first. The salted water might even affect physiological processes. Thus radishes usually contain no starch, but, after treatment with salted water (.3 to 1 per cent. in strength), might contain a great deal. On the other hand, watering cress with 1 per cent. solution caused the starch normally present to disappear, either wholly or in part.

Upon the growth and even on the actual structure of animals, changes of salinity may in some instances exert a marked action. Animals accustomed to develop in fresh water have their growth retarded by the addition of salt. Thus Yung * reared frog's embryos in solutions containing respectively 0, .2, .4, .6, and .8 per cent. of salt, and found that except in the .2 per cent. solution, which had no influence, there was a retardation in development. This increased with the concentration of the solution, so that in the .8 per cent. solution the larvæ took 17 days longer to hatch than in pure water. Again, Sargeant † found that the rate of reproduction by fission of the naid *Dero vaga* becomes slower and slower according to the concentration of the solution it is reared in. Taking the rate in pure water as 11.3, it becomes reduced to 8.5 in .05 per cent. solution of salt, to 7.7 in .1 per cent. solution, 4.1 in .2 per cent solution, and .3 in .3 per cent. solution. Still stronger solutions stop reproduction altogether, and kill off some of the worms.

The more considerable effects which change of salinity may produce are well illustrated by the interesting and widely known observations of Schmankewitsch ‡ upon *Artemia salina* and *A. milhausenii.*

* Arch. des Sci. Phys. et Nat., xiv. p. 502, 1885.

† Davenport's "Experimental Morphology," p. 365.

‡ Zeit. f. wiss. Zoöl., xxv. p. 103, 1875, *and* xxix. p. 429, 1877.

The history of these observations is as follows: Through the breaking of a dam across a salt lake (Kuyalink), a number of individuals of *A. salina* were washed from the upper less saline waters into the lower more concentrated waters, and the Specific Gravity of these was at the same time reduced to 1.058. After the dam was repaired the concentration gradually increased again through evaporation, the Specific Gravity rising to 1.105 the year after; to 1.135 the next year, and to 1.205 the year after that. Accompanying this concentration, the generations of *Artemia* progressively degenerated, till they finally attained the characters of *A. milhausenii*. Schmankewitsch also succeeded in converting a brood of *A. salina* into *A. milhausenii* by the artificial process of gradually increasing the percentage of salt in the water in which they were living (the Specific Gravity being raised from 1.028 to 1.205).

These two forms of *Artemia* have been held to be distinct species, in that *milhausenii* shows an absence of fins and bristles on the lobes of the tail, and has much smaller tail lobes, but larger branchial appendages to the legs, than *salina*. Schmankewitsch himself did not hold this opinion, however, and Bateson,* who has recently studied the question afresh, thinks similarly. Bateson collected samples of *Artemia* from a number of different salt lakes in Western Central Asia and Western Siberia, and, consonant with Schmankewitsch's statement, he found that, on the whole, the number of bristles on the caudal fins, and likewise the size of the fins, was smallest in specimens collected from waters of high Specific Gravity. The accompanying table shows

* "Materials for the Study of Variation," p. 96.

the range in the number of bristles on a single fin, only adult females bearing eggs in the ovisac having been reckoned. Only the Specific Gravities of the various waters are given, but their chemical composition varied between even wider limits, and this may be responsible for some of the irregularities observed:

SP. G.	BRISTLES.	SP. G.	BRISTLES.	SP. G.	BRISTLES
1.030	10—24	1.100	4—10	1.160	16—19
1.050	11—13	1.105	5—9	1.165	1—3
1.056	9—17	1.105	4—8	1.165	1—5
1.065	2—7	1.115	1—6	1.170	6—8
1.075	8—13	1.115	5—9	1.175	1—5
1.075	5—7	1.130	12—16	1.179	4—9
1.085	13—15	1.140	3—7	1.204	2—5
1.095	20—28	1.150	4—10	1.215	2—4
1.100	8—14	1.150	7—8	1.215	2—7
1.100	8—12	1.150	0—1		

As this table and also Schmankewitsch's results show, there is no true differentiation between *A. salina* and *A. milhausenii*, in that these extreme forms are connected by a continuous series of naturally occurring intermediate forms. As Bateson remarks, "it has never been shown that there is a male *A. milhausenii*, with distinctive sexual characters, and among the Branchiopoda the various sexual characters of the second antennæ in the male are most strikingly distinctive of the several forms."

In addition to the particular character in question, Bateson found that there was a great variation in other characters as well. Thus he says: "Almost each locality has its own pattern of *Artemia*, which differs from those of other localities in shades of colour, in average size, or in robustness, and in the average number of spines on the swimming feet, but none of these differences seem to be especially connected with the degree of salinity." Probably the *Artemia* recently found by R. T. Günther * inhabiting Lake Urmi in Persia in such enormous numbers is only another local variety of *A. salina.* It differs from this species in possessing an incompletely segmented abdomen, in the claspers of the male being larger, and in other characters, but Günther says he is nevertheless inclined to agree with Packard that there is only one well-defined Old World species of *Artemia*, viz., *A. salina.*

Schmankewitsch also changed the salinity in the reverse direction, and gradually diluted the salt water containing some *A. salina* till it finally became perfectly fresh. The Crustaceans, which had gone through several generations during the process, had meanwhile so changed their character that in Schmankewitsch's opinion they now resembled the form of the genus *Branchipus.* Thus the last segment of the post-abdomen became divided into two segments, and Schmankewitsch maintained that this division of the segment is the only structural character really differentiating the genus *Branchipus* from *Artemia.* However, Claus † has shown that there are many other points of difference, and that the division in question is not a structural character of great importance. Also *Branchipus* is distinguished by the sexual characters of its males, which possess no structure in any way similar to the great leaf-like second antennæ shown by the male *Artemia.* We must conclude, therefore, that

* Journ. Linn. Soc. (Zoöl.), xxvii. p. 395.
† Anz. Ak. Wiss. Wien., p. 43, 1886.

though decrease of salinity does produce distinct structural changes, yet Schmankewitsch considerably exaggerated their importance, and deduced from them more than he had any justification for.

In addition to *Artemia,* Schmankewitsch * studied the effect of salinity on several other Crustaceans such as *Daphnia rectirostris, Daphnia magna,* and *Branchipus ferox.* He found that in their case also considerable structural and physiological changes were brought about, the fresh- and salt-water forms differing, in his opinion, by characters usually held to be specific.

Equally interesting evidence as to the effect of gradual increase of salinity has been obtained by Bateson † in the case of the common cockle, *Cardium edule.* This mollusc, together with several others, is present in enormous numbers in the brackish waters of the Aral Sea. The waters of this closed basin have been gradually drying up and receding, but the area left exposed "is not a level tract, but contains three considerable depressions, called respectively Shumish Kul, Jaksi Klich, and Jaman Klich. . . These depressions remained, for a time, as isolated lakes, each containing a separate sample of the fauna of the sea living in it." As they gradually dried up, becoming salter and salter, the character of the shells progressively changed. To determine this change, samples were collected at various levels in the lake areas, and were carefully compared. On the western shore of Shumish Kul there were seven very definite terraces of muddy salt, showing the position of the water at various periods during the gradual drying up. The changes produced con-

* *Ibid.*

† Phil. Trans. 1889, B. p. 297.

sisted in (1) a diminution in the thickness of the shells, this being first apparent in the shells of the third terrace. So marked was this change that the shells of the seventh or lowest terrace were almost horny and semi-transparent, and their weight was not a third that of shells from the first two terraces; (2) a diminution in the size of the beak; (3) a high colouration in the shells. This change occurred almost uniformly, the shells of each terrace being very nearly alike in texture, thickness, and degree of colouration; (4) grooves between the ribs appearing on the inside of the shell as ridges with rectangular faces; (5) a great diminution in the absolute size of the shells on the lowest terrace; (6) an increase in the length (greatest antero-posterior dimension) of the shells in proportion to their breadth, this ranging from the average ratio of 1:.80 in the shells from the first terrace to 1:.725 in shells from the seventh. In Jaksi Klich lake the shells from the lowest and most saline deposit were even more elongated, the ratio of length to breadth being as 1:.68 for samples of smaller shells, and 1:.66 for samples of larger ones. Those from the lowest deposit of Jaman Klich showed about the same degree of elongation as those from the lowest terrace of Shumish Kul.

It was very noticeable that the shells of each sample, whether from a separate lake or only from a particular terrace, resembled each other more closely than they did shells from one of the other lakes, or those from another terrace in the same lake, as at Shumish Kul.

In each of the three lakes mentioned (and also in an entirely distinct locality, the lagoon of Abu Kir near to Alexandria), it was thus found that shells which had lived in very salt water had become like each other in possessing the characters of thinness, high colour, small beaks, ribbing on the inside, and great relative length. "In view of these four instances of similar variations occurring under similar conditions," says Bateson, "it seems almost certain that these conditions are in some way the cause of the variations." In that the variations in the quality, texture, and colour of the shell are found developed to nearly the same degree in all the individuals of successive terraces, Bateson considers they may be fairly supposed to be the direct result of environmental change; but the quality of increased proportional length is not found in all the individuals, and hence may have arisen in some other way, as by Natural Selection of the type best fitted to live in the altered state.

A further proof of Bateson's view is afforded by the fact that when the salinity was altered in the direction of diminution, the characters of the shells were similarly changed in a reverse direction. Thus, as already mentioned, the cockles from the very saline lake of Abu Kir resembled those from the lakes of the Aral Sea, but close to this lake are three small areas of water, the Ramleh lakes, of which the water is now quite fresh (owing to their receiving waste water from the irrigations). One of these lakes contains living cockles, and another the shells of extinct ones. Now in both instances the shells are thick and coarse in texture, and comparatively light-coloured. However, the feature of great proportional length still remains.

Other evidence as to the relation between salinity and structure in molluscs has been obtained by Gibbons * for certain tropical and sub-tropical species of *Littorina*. These organisms are confined to more or less brackish waters, and seem incapable of living in pure salt water. Gibbons says he has "met with three of these species, and in each case they have been distinguished from the truly marine species by the extreme (comparative) thinness of their shells, and by their colouring being richer and more varied; they are also usually more elaborately marked." Thus diminution of salinity seemed to have produced thinness of shell in the species as a whole, but within their own limits it was found that the reverse relation held, and that, as in Bateson's observations, the shells became thinner as the water they lived in became more salt.

In marine animals, as in fresh-water ones, increase of salinity probably tends to diminish the rate of growth. Decrease of salinity, on the other hand, may have the reverse effect, and within certain limits actually *increase* the growth rate. Thus Loeb † determined the rate of regeneration of decapitated hydroid polyps (*Tubularia mesembryanthemum*) placed in sea-water of various degrees of dilution and concentration, seven to nine individuals being measured at each concentration. His results are reproduced in the figure given below. Here the Specific Gravity of the water is represented along the abscissa, and the amount of regeneration by the height of the ordinates. We see that the maximum rate of regeneration took place in water of Sp. G. 1.025, or in very considerably diluted sea-water (the Sp. G. of this being about 1.038). At this dilution the re-

* Quart. Journ. Conch., i. p. 339.

† "Biological Lectures delivered at Wood's Holl," 1893, p. 46.

generation was more than twice as fast as in normal water, but with further dilution it rapidly diminished, and ceased altogether in water of Sp. G. 1.013.

Somewhat similar results to these were obtained by the author for sea-urchin plutei. As we saw in the last chapter, the actual size of an organism is probably affected by environment in a similar manner to the growth rate, and the author found that these plutei,

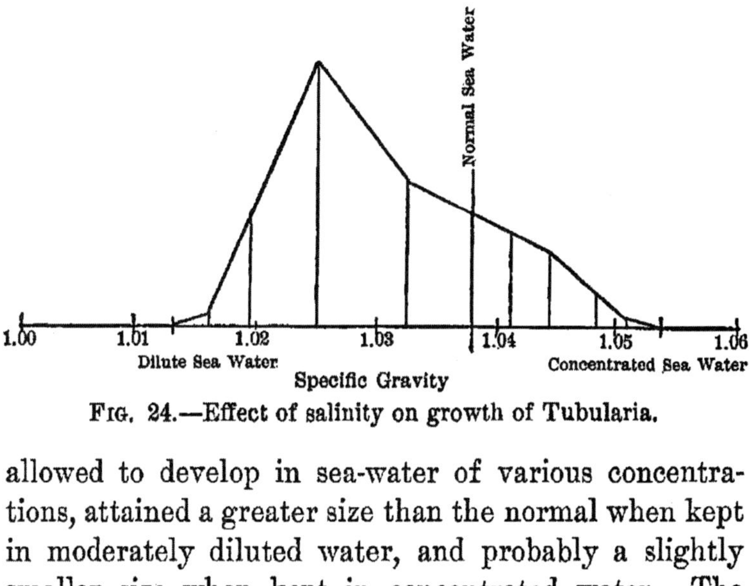

FIG. 24.—Effect of salinity on growth of Tubularia.

allowed to develop in sea-water of various concentrations, attained a greater size than the normal when kept in moderately diluted water, and probably a slightly smaller size when kept in concentrated water. The results obtained are indicated in the subjoined figure, where the abscissæ represent the salinity of the water, and the ordinates the average percentage variation in the size of the larvæ after eight days' growth, as compared with that of larvæ grown in normal sea-water. The salinity of normal water was taken as 1000, and the less saline waters were

* Phil. Trans. 1895, B. p. 586.

obtained by diluting respectively 950, 900 c. c., etc., of water to a litre; the more saline by concentrating 1050, 1100 c. c., etc., to a litre. The maximum effect on size was produced by a solution containing 50 c. c. of fresh water per litre, the increase amounting to 15 per cent. With further dilution, the favourable influence became less and less, till, with water con-

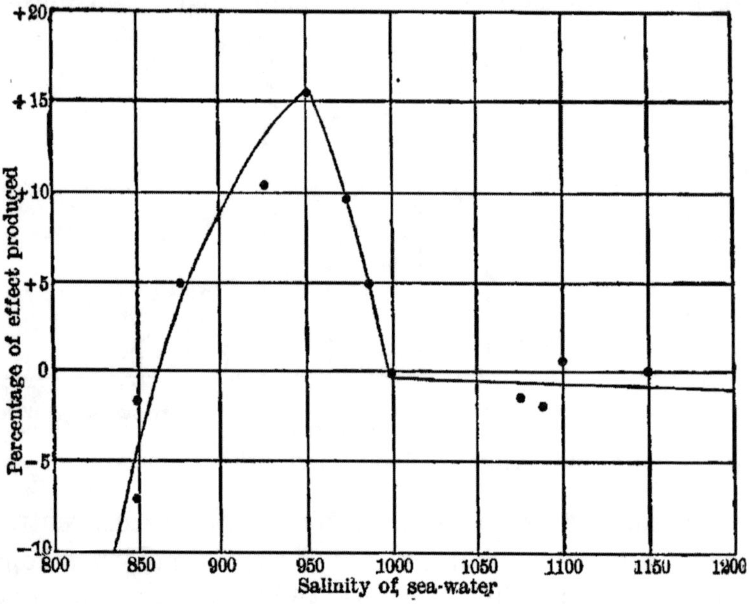

FIG. 25.—Effect of salinity on size of sea-urchin larvæ.

taining 150 c. c. per litre, it was negative. Thus the optimum salinity is for a much less diluted water than in the case of the tubularians. The present results also differ from these latter in that more concentrated waters have exceedingly little effect on the size of the larvæ.

CHAPTER IX.

THE EFFECT OF FOOD AND OF PRODUCTS OF METABOLISM.

Effect of artificial manures on growth of crops—Effect of nutrition on plant variation—Development of bees and of aphides in relation to food—Influence of nature of food on wing markings of certain Lepidoptera—Dependence of colour of larvæ on plant pigments—Influence of food on growth of tadpoles—Plumage of certain birds altered by abnormal diet—Quality of food influences organs of digestion—Every organism probably has specific metabolism, which has especially adverse action on its own growth—Products of metabolism may stimulate growth—Effects of small quantities of urea, uric acid, and ammonium salts—Influence of volume and of surface area of water on growth of pond snail—Influence of surface area on growth of tadpole—Effects of increasing quantities of metabolic products on characters of a snail, and of a Crustacean.

DARWIN records * that Andrew Knight was of the opinion that "of all the causes which induce variability, excess of food, whether or not changed in nature, is probably the most powerful." Darwin himself, moreover, was inclined to accept this view of the potency of food as probable. That changes in the amount and the quality of the food available for an organism during its growth must of necessity exert an important influence on the course of that growth, and presumably, therefore, on the final limits of its attainment, is sufficiently obvious both from one's own everyday experience, and

* "Animals and Plants," vol. ii. p. 244.

from a simple recognition of the relation between cause and effect. Growth can only take place at the expense of food material, and unless this is always more than sufficient for the needs of the organism, the rate of growth must be dependent upon it.

In spite of the importance of changes in feeding as a source of variation, the number of direct and unequivocal experiments made upon the subject is comparatively small, for most of them are complicated by simultaneous changes in other conditions as well. Upon members of the vegetable kingdom, the experiments made by Lawes and Gilbert * at Rothampsted during the last fifty years afford most valuable evidence. These concern the effect of various manures on the growth of barley, wheat, and various leguminous plants. In the accompanying table are given the average

ADDITIONS TO SOIL.	NO NITROGENOUS MANURE.	200 LB. AMMONIUM SALTS, = 43 LB. NITROGEN.	275 LB. SODIUM NITRATE, = 43 LB. NITROGEN.	1000 LB. RAPE CAKE = 49 LB. NITROGEN.
Without mineral manure,	16.5	29	32.7	41.2
Superphosphate,	21.7	42.7	45.7	43.4
Potassium, sodium and magnesium sulphates,	18	31.4	33.5	39.5
Superphosphate and K. Na. and Mg. sulphates,	22.4	43.5	45.5	43.2

amounts of barley grain (in bushels per acre) obtained each year from soils treated in various ways. These observations were carried on for forty years in succession (1852-91) upon the same land, and so represent strictly average results, from which errors due to varia-

* "The Rothampsted Experiments," p. 78, Edinburgh and London, 1895.

tions of season, and other causes, are practically eliminated.

From this table we see that the average yield from land left entirely without manure was 16.5 bushels of grain. On adding various manures, all of which contained about the same weight of combined nitrogen, the yield of grain was doubled, or, in the case of rape-cake manure, increased to two and a half times the amount. The addition of various inorganic salts to the soil also had a favourable effect, though to nothing like the same degree as that of the nitrogenous manures. Thus we see that, when no nitrogenous manure whatever is present, the addition of superphosphates increases the yield by 32 per cent.; of potassium, sodium and magnesium sulphates by 9 per cent.; and of both superphosphates and these sulphates, by 36 per cent. When the nitrogen is added as ammonium salts or nitrates, then combinations of nitrogenous and mineral manures give a very much better yield than the nitrogenous manure alone, but when it is added as rape cake, the growth of the crop has already been so much increased that the further addition of mineral salts effects but little. Yet even the highest of the numbers in this table does not represent the maximum amount of growth of which the barley is capable, for a soil treated with farm-house manure, and no additional mineral salts, yielded on an average 48.6 bushels per acre. In all these experiments the yield of straw was increased in more or less similar proportions to the yield of grain, and hence we may conclude that the growth of a plant in normal soil can be very nearly trebled if only favourable enough conditions are afforded it.

Somewhat similar results to these were obtained by Lawes and Gilbert for wheat, bean, clover, and other crops, but it is deemed unnecessary to reproduce them here.

Most striking evidence as to the influence of nutrition on variations has been obtained by De Vries.* When carrying out his artificial selection experiments on five-leaved clover, he found that in one series of observations seeds from some plants grown in a poor soil yielded 39 per cent. of 3-leaved, and 48 per cent. of 5- to 7-leaved clover. Those from plants of the same stock which had been grown in a rich soil, however, gave only 14 per cent. of 3-leaved, and 73 per cent. of 5- to 7-leaved clover.

The effect of nutrition on *Ranunculus bulbosus* (Crowfoot) was almost as striking.† Wild flowers collected near Hilversum were found by De Vries to have the following frequencies of distribution in the numbers of their petals:

Number of petals,	5	6	7	8	9	10	11
	657	41	11	2	4	0	2

In the autumn of 1887 De Vries planted some of these plants in his culture garden, where they bloomed the following year. Owing, presumably, to the better nutrition, the proportion of flowers with more than five petals was considerably increased:

Number of petals,	5	6	7	8	9	10
	133	55	23	7	2	2

The seed from the many-petalled flowers was collected

*" Die Mutationstheorie," p. 448.

† Ber. d. deutsch. Bot. Ges., xii. p. 197, 1894.

and sown through two seasons, and from the seed then obtained 372 plants were grown. Some of these germinated early, and so developed under less favourable conditions than the others. As will be seen from the accompanying figures, these early plants had 9 petalled flowers occurring the most frequently, whilst the later ones had 10 petalled flowers; *i. e.*, flowers with twice the original number of petals:

Number of petals,	5	6	7	8	9	10	11	12	13	14	15	16—31
Early plants,	409	532	638	690	**764**	599	414	212	80	29	18	20
Late plants,	40	52	126	165	204	**215**	177	104	35	8	4	0

The somewhat unexpected results obtained by MacLeod * with *Ficaria ranunculoides* may also be attributed, at least in some degree, to the effects of nutrition. MacLeod determined the numbers of stamens and of pistils in the flowers borne by a number of plants at the beginning of the flowering season, and again in the flowers borne by the *same* plants at the end of the season. The early flowers had on an average 26.73 stamens and 17.45 pistils, whilst the late ones had only 17.86 stamens, and 12.15 pistils. Also the variability in the number of stamens and of pistils was very different in the two cases, the coefficients of variation being respectively 14.1 and 22.3 per cent. in the early flowers, and 18.5 and 27.9 per cent. in the late ones. The method adopted by MacLeod for estimating the correlation between the numbers of stamens and of pistils is erroneous, so Professor Weldon has recalculated the constants.† He finds that MacLeod's figures indicate the correlation to be much less in the early than in the

* Botanisch. Jaarboek., xi., 1897.

† Biometrika, i. p. 125, 1901.

late flowers (the *r* being respectively .51 and .75 in the two cases). As Professor Weldon remarks, these results "provide a most valuable lesson as to the possible danger of asserting that such differences are significant of local races."

By observations upon the growth of seedlings placed in various solutions, it has long been known that normal growth is possible only if various inorganic salts are present. There must be nitrogen in the form of nitrates or ammonium salts, sulphur in the form of sulphates, phosphorus as phosphates, chlorine as chlorides, and the metals sodium, potassium, magnesium, calcium, and iron in solution as salts. The absence of any one of these substances speedily inhibits normal growth; as soon, in fact, as the seedling has exhausted the small quantity of it stored up within itself. For instance, plants grown in solutions containing no iron soon show a sickly appearance; "the leaves are no longer green, but white, and microscopic examination of them shows that abnormal chlorophyll bodies, or none at all, are present in their cells. If we add to the food solution a few drops of dilute ferric chloride solution, the previously white leaves become green in two or three days, and the growth of the plants now proceeds normally." * It follows, therefore, that if the absence of these various substances stops growth altogether, a deficiency in them must produce diminished or abnormal growth, and so lead to the production of variations.

With members of the Animal Kingdom, variations in the inorganic salts of the food may also be a source of

* Quoted from Detmer's "Practical Plant Physiology," p. 84.

variation. Thus Cooke * states that "a deficiency of lime in the composition of the soil of any particular locality produces very marked effects upon the Mollusca which inhabit it; they become small and very thin, occasionally almost transparent. The well-known var. *tenuis* of *Helix aspersa* occurs on downs in the Channel Islands where calcareous material is scarce. For similar reasons, *H. arbustorum* develops a var. *fusca*, which is depressed, very thin, and transparent, at Scilly and also at Lunna I., E. Zetland."

However, in animal development the supply of inorganic salts is almost always more than sufficient for the needs of the organism, and such variations as are produced are due chiefly to the organic constituents of the food. Among invertebrate animals, our knowledge of the direct influence of food is almost confined to certain of the Insecta. In the case of bees, it has been known for a very long time that the quality and quantity of the food supplied to the larvæ determines whether the reproductive organs shall undergo their full development, and produce fertile queens, or remain undeveloped, and so produce non-fertile working females. According to A. von Planta, the diet of the queen larvæ contains twice as much fatty material as that of the workers.† Again, Eimer has pointed out that in the case of the humble bee, the first brood of ova, laid in the spring, get only a scanty supply of nutriment, and develop into small females, which are fertile though they can only produce drones. The next brood

* Vol. iii., "Cambridge Natural History," p. 89.

† Quoted from Geddes and Thomson's "Evolution of Sex," p. 43, 1889.

born obtain more nourishment, and develop into larger females, which are capable of occasionally producing females, as well as drones. Finally the future queens, which obtain a still richer diet, are born. The determination of sex seems to be dependent on nutrition also in aphides or plant-lice. Thus "during the summer months, with favourable temperature and abundant food, the aphides produce parthenogenetically generation after generation of females. The advent of autumn, however, with its attendant cold and scarcity of food, brings about the birth of males, and the consequent recurrence of strictly sexual reproduction." * In this instance, therefore, the effect of nutrition is bound up with that of temperature, and there are no data to show whether either of these conditions could produce the effect if acting alone.

Upon the Lepidoptera the effects of various foods have been tested in a considerable number of instances. Observations were made by G. Koch † in Germany as long ago as 1832. By feeding the caterpillars of *Chelonia hebe* with different plants, he obtained specimens which were either fiery or dull red on the under wings, and which varied in the extent of black marking and white ground. In the case of *Euprepia caja* (Common tiger moth) it is known, Koch says, "that when the caterpillars are fed from their hatching to their metamorphosis with leaves of lettuce or deadly nightshade, not one of the imagines produced resembles the original form; when the insects have been fed on lettuce, the white ground-colour of the wings predominates;

* *Ibid.*, p. 46.

† Eimer, "Organic Evolution," p. 149.

when fed on deadly nightshade the brown markings of the upper wings often coalesce and the white vanishes; in like manner the blue markings on the lower wings fuse together and displace the orange-yellow ground colour." *

A careful series of observations upon various moths, extending over some ten years, has been made by Gregson.† His results may be tabulated as follows:

Pygæra bucephala (Buff Tip) is finer and darker when fed upon sycamore.

Xylophasia polyodon (Dark Arches) is dark, sometimes black, when fed upon heather.

Hadena adusta (Dark Brocade) is darker when fed upon heath.

Acronycta menanthydis (Light Knot-Grass) when fed on sallow, often produces var. *A. salycis*; fed on heath, produces light specimens.

Hybernia defoliaria (Mottled Umber) is beautifully marked when fed upon birch; but on elm gives dull-coloured forms, almost without markings.

Eupithecia venosaria (Netted Pug) fed on inflated catchfly is almost white; on shore catchfly is much larger and almost lead colour.

Noctua festiva (Engrailed Clay) fed on thorn is rich red and well marked: on grasses is light yellowish, and rarely well marked.

Noctua triangulum (Double-Square Spot) fed on thorn is dark: fed on low plants is light.

Abraxas grossulariata (Magpie) fed on red currant is light; on blackthorn is darker; on bullace or wild plum is darker still, the white sometimes becoming yellow.

The following case, recorded by the late Mr. Newman,‡ is of especial interest in that it occurred under natural conditions. The larva of *V. polychloros* (Large

* *Ibid.*, p. 151.

† The Zoölogist, p. 7903, 1862.

‡ The Entomologist, vi. p. 88, 1872.

Tortoiseshell) usually feeds upon elm, but that of *V. urticæ* (Small Tortoiseshell) upon nettles. Some larvæ were found by Mr. J. A. Tawell feeding upon nettles, and so were considered to be those of *V. urticæ*. They were accordingly kept on nettles, but to his surprise developed into *V. polychloros* imagines. "These specimens," records Newman, to whom they were shown, "have a wonderful similarity to *urticæ*, which they do not at all exceed in size; still the colour is nearer to that of *polychloros* than that of *urticæ*." The effect of abnormal food on *Melitæa artemis* (Greasy Fritillary) has been noticed by H. Goss.* By feeding the larvæ on honeysuckle, a series of very dark imagines was obtained, which differed both in size and colouring from all other specimens known to him, though these had been derived from very varied localities in England, Ireland, and Scotland. Again, the quantity of the food supplied may have as considerable an effect as the quality. By mistake some *V. io* (Peacock butterfly) larvæ, captured by Mr. R. Cox,† were left for several days without fresh food, and all the dead leaves and stalks were devoured. Nearly all the imagines obtained from them were rather small, but they also varied much in the intensity of their colouring, two specimens being very much darker than usual, with the yellow in the costal spot and ocellus much reduced. It seems to me, however, that probably these changes were due rather to the abnormal food devoured by the larvæ than to the actual lack of food.

As regards the larvæ of Lepidoptera, the obvious re-

* The Entomologist, vii. p. 203, 1874.

† The Entomologist, ix. p. 58.

lation between their colour and that of their food seems to show that the one is directly dependent on the other. Meldola * accounted for it by supposing that the larvæ had been rendered transparent by Natural Selection, whereby the colour of the vegetable food eaten was itself enabled to give the colour to the larvæ. Poulton † has shown that the colours of the larvæ are due partly to the pigments proper to the larva, and partly to the pigments derived from the food plants. These pigments undergo some modification in the tissues, but Poulton states that as far as he has investigated the subject "all green colouration without exception is due to chlorophyll; while nearly all yellows are due to xanthophyll." The chlorophyll, or some modification of it, tinges the blood of the larvæ, the green colour of which is often due to this cause alone.

From these observations, therefore, it follows that a change of food may also effect a change of colouration. That this is so is strikingly shown by some other observations by Poulton.‡ Obtaining a large number of larvæ of *Tryphæna pronuba* (Common yellow underwing) from the same batch of eggs, he split them up into three groups. One he fed on the white midribs of the cabbage, from which the yellow blade had been carefully removed with scissors. These larvæ remained almost white at first, and afterwards showed a moderate amount of black pigmentation. The other two groups of larvæ he fed respectively on the yellow etiolated leaves from the heart of the cabbage, and upon the deep

* Proc. Zoöl. Soc. 1873, p. 155.

† Proc. Roy. Soc., xxxviii. p. 269, 1885.

‡ Proc. Roy. Soc., liv. p. 417, 1893.

green external leaves. These larvæ, however, were all of a bright green or brown colour. Hence it would seem that both etiolin and chlorophyll are capable of being transformed into a larval colouring matter, which may be either green or brown.

As regards the effects of feeding among vertebrate animals, a careful series of experiments upon tadpoles (*Rana esculenta*) has been made by Yung.* The tadpoles were all derived from the same batch of eggs, and were placed, in groups of fifty, in six similar jars of water. All the conditions of development such as light, temperature, and frequency of change of water, were identical, the food alone being varied. The kinds of food supplied, and the average size attained by the tadpoles after 42 days' development (three being measured in each case), are given in the accompanying table:

NATURE OF FOOD.	AQUATIC PLANTS ANACHARIS CANADENSIS AND SPIROGYRA.	FIRST ALBUMINOUS ENVELOPE OF FROGS' EGGS; THEN LIQUID ALBUMEN OF HEN'S EGG.	PIECES OF YOLK OF HEN'S EGG.	PIECES OF COAGULATED ALBUMEN OF HEN'S EGG.	PIECES OF FISH FLESH (MINNOWS).	PIECES OF BEEF.
Length of tadpole Breadth of tadpole	18.3 4.2	23.2 5.0	26.0 5.8	33.0 6.6	38.0 8.8	43.5 9.2
Per cent. of frogs after 58 days.	0	0	14	20	48	66

Here we see that the purely vegetable diet acted least favourably, and the beef diet the most favourably. Egg yolk did not answer so well as coagulated egg albumen, but better than uncoagulated albumen. From

* Arch. de Zoölogie Expér., 1883, p. 3.

the bottom line of the table we see that, 58 days after the beginning of the experiment, none of the 50 tadpoles fed on plants and on liquid egg albumen were surviving; but of those fed on fish and on beef, respectively 48 and 66 per cent. were alive, and had undergone their metamorphosis into frogs.

The effects of certain foods on the plumage of birds is well known to bird fanciers. Thus hemp seed causes bull-finches and certain other birds to become black. Cayenne pepper, mixed with the food, changes the yellow colour to an orange red. This colour change can only be effected by feeding the very young birds; with adults there is no effect whatever. Sauermann* found that all races are not equally susceptible to the abnormal diet, some being changed to a crimson, others to a beautiful orange, whilst others remain absolutely unaffected. He found also that canaries are not alone in their susceptibility, for on feeding some white Italian fowls, eight weeks old, with the pepper, orange stripes appeared on the breast feathers of one of them after ten days. Later on, the whole body was covered with mixed white and orange feathers, and the breast had become red. One other fowl also developed a red breast, but the remaining ten showed no change whatever. The doses of Cayenne pepper given were enormous (50 gm. daily), so that the conditions were absolutely unnatural.

More remarkable than these observations are the facts ascertained by A. R. Wallace, and communicated by him to Darwin.† Thus he states that "the natives

* Archiv f. Anatomie u. physiol. Physiol. Abtheil., p. 543, 1889.
† "Animals and Plants," ii. p. 269.

of the Amazonian region feed the common green parrot (*Chrysotis festiva*) with the fat of large Siluroid fishes, and the birds thus treated become beautifully variegated with red and yellow feathers. In the Malayan archipelago, the natives of Gilolo alter in an analogous manner the colours of another parrot, namely, the *Lorius garrulus*, and thus produce the *Lori rajah* or King Lory."

As regards mammals, it is asserted by Nathusius * that if rich and abundant food be supplied to young pigs, it has the direct effect of producing a broader and shorter head. Poor food, on the contrary, produces a longer and narrower head, or a tendency towards the characters of the wild boar. Again, Krocker † has shown that the amount of wool yielded by sheep is greatly influenced by the quantity of food. The following are the weights of wool yielded per day by sheep weighing in aggregate 1000 kilograms:

DIET.	KILOGRAMS OF WOOL.
Scanty winter food,	.69
Plenty of hay,	.87
Good pasture,	.96
Fattening process,	1.08—1.24

The quality of the food may considerably affect the organs of digestion. Thus Cuvier ‡ found that in the wild boar the length of the intestines is to that of the body as 9 to 1, but in the common domestic boar it is as 13.5 to 1. It is, of course, impossible to say for certain whether this increased length was the direct result of a more vegetable diet, but it seems highly probable that

* "Schweineschädel," p. 99; *also* "Animals and Plants," i. p. 75.
† De Varigny's "Experimental Evolution," p. 90.
‡ "Animals and Plants," i. p. 77.

this was so, at least in part. The observations which have been made from time to time as to the effects of various kinds of food on the thickness of the stomach wall, are, however, free from all such doubt. The change produced must evidently be the direct result of the altered diet. Thus John Hunter observed a most marked thickening and hardening in the stomach of a gull (*Larus tridactylus*) which had been fed for a year on grain. It is stated by Dr. Edmondston that a similar change takes place under natural conditions every year in the stomach of the common Herring gull (*Larus argentatus*). Thus in the Shetland Islands this bird feeds in the winter on fish, but in the summer frequents the cornfields and feeds on grain. Dr. Edmondston has also noticed a somewhat similar change in the stomach of a raven which had been fed for a long time on vegetable food. Again, Menetries found that in an owl (*Strix grallaria*) the effect of vegetable diet was to change the form of the stomach, and make the inner coat leathery.*

The converse experiment of feeding graminivorous birds on a flesh diet has been made by Dr. Holmgren. By feeding pigeons on meat for a considerable time, he found that the gizzard gradually acquired the qualities of a carnivorous stomach. Again, Delage † fed a fowl for three years on meat, and found that the muscular substance of its gizzard was considerably decreased. All these results, though apparently so unequivocal, have not passed unchallenged; for G. Brandes,‡ who

* *Vide* "Animals and Plants," ii. p. 292.
† L'Année Biologique, 1896, p. 468.
‡ Biol. Centralblatt, xvi. p. 825.

fed both flesh-feeding birds on grain, and grain-feeders on flesh, states that he was unable to trace any adaptation to the altered conditions in either case.

The Effects of Products of Metabolism. That organisms react on each other has long been recognised. The interdependence is especially obvious in the case of parasite and host; but reflection will show, I think, that the interaction is of much wider scope than is included in such self-evident cases as these. In any given volume of water, or any given area of land, every animal and every vegetable organism may to some extent affect the well-being of every other organism, both animal and vegetable. The animal does this largely through the agency of its own specific metabolism, or through the specific products of excretion which, coming into contact with the other organisms, in turn affect them. That every species of animal does possess a specific metabolism is, perhaps, scarcely what one would on *a priori* grounds expect; but the observations made by the author * tend to prove that such is actually the case. These observations chiefly concern Echinoids, both adult forms and plutei, but more especially the already so frequently mentioned plutei of *Strongylocentrotus.*

On allowing the fertilised ova of *Strongylocentrotus* or of *Echinus microtuberculatus* to develop in water in which another batch of larvæ (*Strongylocentrotus, Sphærechinus* or *Echinus*) had already been developing for 8 to 12 days, but from which they had been removed by filtration, it was found that in every case they

* *Vide* Mittheilungen a. d. Zool. Stat. z. Neapel., Bd. xiii. p. 389 *et seq.*

were diminished in size. In ten experiments the average diminution was 7.1 per cent. It was concluded, therefore, that the first batch of larvæ had excreted some products of metabolism into the water which had adversely affected the growth of the second batch. Other observations * showed that the growth of larvæ may be affected by their own metabolic products. Thus it was found that the arm lengths of the larvæ became smaller and smaller the larger the number of larvæ allowed to develop together in a given volume of water. In the accompanying table are given the mean results of 159 sets of measurements, each on the anal and oral arm lengths of 50 larvæ.

NUMBER OF DIFFERENT OBSERVATIONS.	NUMBER OF LARVÆ PER LITRE.	MEAN LENGTH OF ANAL ARM.	MEAN LENGTH OF ORAL ARM.
37	Under 1500	121.2	118.4
32	1500 to 3500	114.0	110.5
21	3500 to 6000	105.8	101.0
34	6000 to 11,000	102.9	99.4
27	11,000 to 20,000	95.7	94.2
6	20,000 to 30,000	85.5	86.3
2	Over 30,000	56.6	68.5

Here we see that when less than 1500 larvæ were developing together, their relative anal and oral arm lengths were respectively 121.2 and 118.4. As the number increased, the lengths steadily dwindled down, till with over 30,000 per litre they became reduced to respectively 56.6 and 68.5, or about half their original amount. Now it was found that the body lengths of the larvæ,

* Phil. Trans. 1895, B. p. 603.

or the dimension measured in all the observations on larvæ hitherto described, was practically unaffected by the "concentration" of the larvæ. This apparent contradiction is easily accounted for by the fact that the times of development of the body and of the arms of the larvæ is not the same. At moderate temperatures, the body attains about 80 per cent. of its full length by the end of the second day, and 90 per cent. by the end of the third. The arms are practically non-existent at the end of the second day, however, and attain only 65 per cent. of their full length by the end of the third. As, therefore, the products of metabolism in the water are practically nil during the first day or two, and only gradually accumulate with progress of time, it follows that the growth of the body tissues is unaffected by them, whilst that of the arm tissues is restrained.

The influence of the excreta of adult Echinoids upon larval growth was then tested. Echinoids of known weight were kept for a known time in a known volume of water, so that, on determining the absolute effect produced on larvæ grown in this water, it was possible to calculate the relative effect produced by unit weight of Echinoid kept for unit time in unit volume of water. On growing larvæ in water previously fouled by adult Echinoids of their own species, it was found that, as a mean of five observations, they were diminished in relative size by 2.6 per cent., whilst only 41 per cent. of the ova employed reached the larval stage. On growing them in water fouled by Echinoids of other than their own species, the larvæ, as a mean of five observations, were diminished by only 1.9 per cent., whilst 54 per cent. of the ova reached the larval stage. That is to say, the products of excretion of an Echinoid act more adversely both on the death rate and on the growth of embryos if these belong to its own species, than if they belong to another species. At least this is the case with *Strongylocentrotus*, *Sphærechinus*, and *Echinus*. With two other (physiologically) less closely related species, viz., *Arbacia pustulosa* and *Dorocidaris papillata*, it was even found that the products of excretion, so far from acting adversely on growth, actually favoured it. Thus *Strongylocentrotus* larvæ grown in water fouled by these two species were increased in size by respectively 4.3 and 1.7 per cent., whilst respectively 81 and 50 per cent. of the ova employed reached the pluteus stage.

It will probably be thought that this last result is erroneous; but other observations showed that it was not so. Thus *Strongylocentrotus* larvæ were grown in water fouled by various other animals, and it was found that in this case also there was generally a distinct increase in size. We see in the accompanying table that

ANIMALS USED FOR FOULING WATER.	ABSOLUTE PER CENT. VARIATION IN SIZE OF LARVÆ.	RELATIVE PER CENT. VARIATION	ANIMALS USED FOR FOULING WATER.	ABSOLUTE PER CENT. VARIATION IN SIZE OF LARVÆ.	RELATIVE PER CENT. VARIATION.
1 Fish	+1.4	+1.8	3 Holothurians	+5.0	+2.0
2 Fish	+8.3	+12.8	3 Holothurians	+4.7	+.9
4 Crabs	+1.6	+2.1	1 Crab + 3 Anemones	−1.5	−1.0
30 Molluscs	+2.8	+1.3	3 Anemones	−.6	−.5
48 Molluscs	+4.8	+1.1	1 Medusa	−2.2	−1.9

of the ten observations made, a positive effect (averaging 4.1 per cent.) was produced in seven instances, whilst a much slighter negative effect (averaging 1.4 per cent.) was produced in only three. The relative variation in size produced by 100 grams of animal fouling 1 litre of water for 1 hour is also given. These values are somewhat more variable than those representing the absolute variation, but they to some extent corresponded to the amount of nitrogenous matter actually excreted into the water, as was proved by chemical analysis of the various samples.

We may conclude, therefore, that under certain conditions products of metabolism may stimulate an organism to increased growth, whilst under certain others they may retard growth. What is the nature of these excretory products which exert so potent an effect? Observations made on the influence of various simple substances on larval growth seem to throw some light on the question. The results obtained with uric acid and urea are given in the accompanying table:

SUBSTANCE PRESENT, AND AMOUNT.	PER CENT. VARIATION IN SIZE OF LARVÆ.
Uric acid, 1 in 154,000	+ 5.3
" " 1 in 70,400	+12.2
" " 1 in 58,000	+ 5.8
" " 1 in 28,000	− 2.1
Urea 1 in 65,000	+ 2.3
" 1 in 59,000	+ 3.7
" 1 in 44,000	+ 2.2

Here we see that uric acid in moderate amounts exerts a very favourable influence on the size of larvæ. It is only when the proportion is raised to 1 in 28,000 (a more than half saturated solution), that an unfavourable effect shows itself. Urea also acts favourably, though not to the same extent as uric acid. If, therefore, these two simple bodies are capable of stimulating the tissues to increased growth, it is possible that the effects produced by animal excreta may be due to minute quantities of other but more complex nitrogenous bodies. That they are not due to simple urea and uric acid was proved by the chemical analyses of the fouled waters, for the amount of nitrogen found to be present was never half sufficient, and as a rule was very much less. As to the substances producing an adverse influence on growth, no definite evidence was obtained, but it seemed possible that they might be derivatives of ammonia, perhaps amines or amido-bodies. Thus ammonium salts themselves exert an exceedingly poisonous action, as may be gathered from the following data:

WEIGHT OF AMMONIUM CHLORIDE PER LITRE.	EFFECT PRODUCED.
.0258 gm.	Larvæ diminished 7.3 per cent. in size.
.0394	" " 19.0 per cent. "
.1075	59 per cent. blastulæ formed. Larvæ lived 3 days.
.3745	37 per cent. " " No larvæ.
.7890	Most of the ova had disintegrated after 24 hours.

That the effect produced by nitrogenous bodies depends almost entirely upon the form in which the nitrogen is combined, is shown by the fact that nitrates and nitrites have no influence on larval growth unless the proportions added be over 1 gram and .3 gram per litre respectively.

The products which every organism excretes probably consist, therefore, of various complex nitrogenous bodies, which differ in different organisms. If they come into contact again with the tissues from which they have been expelled, they retard the growth of these tissues, but if with other tissues with which they have no direct chemical relation or association, they may under certain circumstances stimulate them to increased growth.

The effects of products of metabolism upon growth have been tested at considerable length in the case of certain Molluscs. At least it is to this influence that the results obtained by Karl Semper and by De Varigny in their experiments on *Limnæa stagnalis,* the common pond snail, ought, in my opinion, to be ascribed. Semper * found that if various numbers of the small snails were placed in equal volumes of water immediately after hatching, and were kept there under otherwise equal conditions as to food, temperature, etc., for about two months, then the size to which they attained was by no means equal, but varied in more or less inverse proportion to the number of snails present. In four very consistent experiments, the numbers of snails placed in volumes of 2000 cc. of water were in each case respectively 1, 5, 10, and 20, or each snail obtained respectively 2000, 400, 200, and 100 cc. of water. The lengths attained by the snails after two months' growth are given in the table below.

Here we see that snails allowed to grow singly in the 2000 cc. vessels of water attained to more than three times the size of those grown in twenties. This was not merely a question of nutrition, as the amount of food

* Arb. a. d. Zool. Inst. in Würzburg, i. p. 137, 1874; *also* "Animal Life," ed. 4, p. 51.

supplied was always at an optimum. It was evidently in some way the result of the *volume* of water available for each snail's needs. Other experiments in which the number of snails was constant, but the volumes of water unequal, gave a similar result. The manner in which the volume of water affected the snail's growth, Semper confessed himself unable to determine; but he supposed that the water must contain some substance, as yet unknown, which is essential for stimulating the growth of the snails. The less of this hypothetical body available, therefore, the more retarded their growth.

NUMBER OF SNAILS IN 2000 CC.	VOLUME OF WATER PER SNAIL.	LENGTH IN MILLIMETRES.				AVERAGE LENGTH.
1	2000 cc.	17.5	19.7	18.5	17.0	18.2 mm.
5	400 cc.	11.7	10.1	10.8	10.5	10.8
10	200 cc.	8.8	7.5	6.8	8.6	7.9
20	100 cc.	6.2	6.2	4.6	5.0	5.5

Within recent years De Varigny * has re-studied the unsolved problem, and has extended Semper's methods in several directions. He used both *Limnæa stagnalis* and what he termed *L. auricularis*, though this form was probably *L. pereger*, judging from his figures.† He confirmed Semper's conclusion that the size is influenced by the number of individuals in the vessel, but he did not find the snails nearly so sensitive to differences in the volume of the water as had Semper. Differences in the superficial area of the water exposed to

* Journ. de l'Anat. et de la Physiol., p. 147, 1894.

† Nat. Sci. v. p. 168.

the atmosphere he found to be much more important than differences of volume. Thus a snail kept five months in a litre of water having a surface of 18 cm. in diameter attained to nearly twice the length of one kept in an equal volume of water which had a surface of only 2 cm. diameter. In order to test Semper's hypothesis of the essential substance in the water, De Varigny suspended a glass tube 2 to 3 cm. in diameter in various sized vessels of water. A piece of muslin was tied over the bottom of the tube, so as to permit of interchange of water, but prevent the snails placed in the tube and in the outer vessel of water from intermigrating. After two to five months' growth it was found that the snail placed in an outer vessel of 4200 cc. capacity sometimes attained to more than twice the length of that placed in an inner one of 250 cc. capacity. Again, snails were placed in two tubes of the same size, one of which was suspended in a vessel containing 100 cc. of water, and the other in a vessel containing 1150 cc.; in another similar experiment the external volumes of water were respectively 50 and 500 cc. In each case, however, the snails in the two inner vessels attained to practically the same size. Still again, two similar tubes, holding 50 to 70 cc. of water, were placed in a vessel containing 4200 cc. of water. One tube was closed with muslin, and the other with a tight-fitting cork, which of course prevented all interchange with the outer vessel of water. Nevertheless the snail in this tube, after two months' growth, was only very slightly smaller than that in the other tube, but both of them were only about three-fifths the size of the shell grown in the external vessel. It should be mentioned that in all these experiments De Varigny lifted each tube out of its vessel of water and replaced it two or three times a day, in order to mix the water in it with that in the external vessel. He concluded, therefore, that Semper's hypothesis is not tenable, and that the size of the snails actually depends in some way on the volume of water containing them, and on the superficial area of this water. His explanation of the phenomenon is that in small vessels the snail would need to move about less in order to obtain food, for this would always be near at hand. With less exercise, the growth rate might accordingly be diminished. This explanation does not account for some of the principal results obtained by Semper and by De Varigny himself, however. Thus in vessels of equal volume, but containing various numbers of snails, the amount of movement and exercise necessary would be just the same in each case, and yet, as we have seen, the growth rate varies enormously.

In all probability, the results obtained both by Semper and by De Varigny can be most simply accounted for in the manner already suggested. Thus De Varigny actually found that snails grown in water in which other snails had already been growing several months were distinctly smaller than those grown in fresh water, and if the excreta of snails had been added as well, they were smaller still. If, then, the observed differences in growth are due to the accumulation of various quantities of products of metabolism, how can we account for the results obtained by De Varigny in his muslin-bottomed tube experiments? We must imagine that the mixing of the internal and external waters two or three times daily, and the constant slow interchange through the muslin, were insufficient to equalise the proportions of metabolic products in the two vessels for more than a short time, so that on an average the water in the inner vessel was more foul than that in the outer. This fouling would probably be much increased by particles of decomposing vegetable matter and of animal excreta collecting in the fibres of the muslin and on the inner walls of the glass tube, and constantly poisoning the water. The outer water would also be fouled in this manner, but to a very much slighter extent, for the "fouling area" of muslin and walls of vessels would be proportionately very much less. That the metabolic products from unhealthy or decomposing vegetable matter can exert a most harmful influence on growth is shown by some of my own experiments with plutei. Thus ova allowed to develop in water which had previously contained 1 or 2 gm. per litre of (presumably unhealthy) seaweed, were diminished in size by as much as 13.2 and 18.1 per cent.*

De Varigny's experiments on the influence of superficial area of water must be considered in conjunction with some observations by Yung † on tadpoles. Yung put twenty-five freshly hatched tadpoles in each of three vessels which contained equal volumes of water (1200 cc.), but of which the diameters were respectively 7 cm., 11 cm., and 14.5 cm. Thus the surface of water exposed to the air varied in the proportions of 1 : 2.5 : 4.3 After a month and a half the tadpoles were found to have attained the following average sizes:

* Mittheilungen a. d. Zool. Stat. z. Neapel., Bd. xiii. p. 348.

† Arch. des Sci. Phys. et Nat., xiv. p. 502, 1885.

SUPERFICIAL AREA OF WATER.	1	2.5	4.3
Length of tadpoles	26.2 mm.	34.2	41.2
Breadth " "	6.1	7.8	8.8
Date of first metamorphosis .	Aug. 4	July 22	June 8

The greater size of the tadpoles bred in the more exposed water Yung attributed to this water absorbing a larger proportion of oxygen from the air. This is in all probability the correct explanation both of these observations and of the similar ones of De Varigny on snails. The greater supply of oxygen would not only stimulate the rate of growth of the tadpoles and of the snails, but would also hasten the oxidation of the harmful products of metabolism. It is true that De Varigny found that a snail kept for eight months in a corked vessel containing about 550 cc. of water and 500 cc. of air attained to only slightly less a size than another snail kept in a similar but unstoppered vessel, but this may have been due to the fact that green plants were flourishing healthily in each vessel throughout the whole period, and these may have been sufficient to remove most of the metabolic products excreted by the snails.

Further evidence as to the influence of volume of water on the growth of molluscs has been obtained by Whitfield.* This observer kept a *Limnæa megasoma* in a small aquarium, and after some months it deposited eggs. These hatched out, grew in size, and in due course themselves deposited eggs. This process continued for four generations in all, the shells of each generation being smaller than those of the one before.

* Bull. Amer. Mus. Nat. Hist., vol. i. p. 29; *and* Amer. Naturalist, xiv. p. 51.

Those of the last generation had altered so much that a conchologist of experience was of the opinion that they could bear no possible specific relation to those of the first. Thus in addition to the diminution in size, the spire had become very slender. In a second experiment of a similar kind, the shells of the third generation were only 4-7ths as long as those of the parent stock, and, still more remarkable, the male organs had disappeared, whilst the liver had become considerably reduced in size.

These extraordinary effects were probably due to the cumulative action of the increasing quantities of metabolic products in the water in which the molluscs were living.

Still another series of observations on the effect of a confined volume of water was described by Warren * only a year or two ago. These were made upon *Daphnia magna* (Water-flea). Four adult individuals were placed separately in covered glass vessels containing 200 cc. of water, together with some Conferva and some mud containing algæ, etc. Four others were placed in similar, but uncovered vessels, and four more in still other vessels, of which the water was changed about once a day. The water in the former vessels was never changed, but the loss due to evaporation in the uncovered vessels was compensated for by the occasional addition of rain water. The Daphnias produced broods of four or five offspring each after about 15 days, and these offspring were allowed to grow in the vessels, and after a time produced offspring in their turn. It was interesting to note, however, that in the vessels in

* Q. J. Microsc. Sci., vol. 43, p. 212, 1900.

which the water remained unchanged, the rate of reproduction and the number of offspring in a brood was considerably diminished. The third generation was produced about 22 days after the second, and the fourth about 25 days after that, and then the breeding stopped. In the vessels with frequently changed water, the third to seventh generations were produced at intervals of respectively 18, 14, 15, 16, and 26 days, and then for some unknown reason the families died out. The confined volume of water had another and even more remarkable effect, however, as it caused a distinct shortening in the length of the spine formed by the posterior prolongation of the carapace. Thus in one series of observations it was reduced from a length of 241 (relative to the carapace length taken as 1000) in the parents to one of 171 in the offspring; and in another series from a length of 276 in the parents to one of 249 in the children, and 185 in the grandchildren. In this latter case, therefore, it would seem as if the acquired character of shortened spine was inherited.

Warren attributes these remarkable effects to the excretory matter thrown off by the Daphnias into the water. Also he concluded that this matter "may feasibly be supposed to be particularly injurious to Daphnia; for when the Daphnia are fast disappearing, there may be a swarm of Ostracods or Copepods (still living healthily in the water)." In fact Warren inclined to the view, already suggested by the author in the case of Echinoids and other marine animals, that water fouled by Daphnia "becomes specifically injurious to Daphnia."

CHAPTER X.

THE EFFECTS OF CONDITIONS OF LIFE IN GENERAL.

Local conditions of life perhaps the cause of local races, but proof of this is as a rule impossible—American and European trees compared—Alpine and Arctic plants—Effects of cultivation—Local races of oysters and of snails—Lepidoptera in Malay Archipelago—Local races of shrimps, of mackerel, and of herring—North American birds and mammals—Action of climate on goats and on rabbits—Effect of domestication on rabbits, pigeons, fowls, and ducks.

In the three preceding chapters we have examined numerous cases of variation produced wholly or in great part by a change in some one condition of environment. In the present chapter no such attempt is made to trace an effect to any single cause, but we shall examine the effects of all conditions of life together, such as climate, nutrition, moisture, and sunlight, in the production of variations. The variations more particularly to be studied are those which are common to whole groups of organisms, and which go to form what are known as local races. Unfortunately in the majority of cases it is impossible to prove that such races are the direct or indirect results of the differences of environment, even when there is a high probability that such is the case. Hence, when local races are referred to, it is not intended to imply that the distinguishing characters exhibited are definitely due to the action of the environment. The inference is only that they *may* be, if not wholly, then in part.

Among plants, a striking instance of the apparently direct action of conditions of life in the production of variations has been afforded by Meehan.* This observer "has compared twenty-nine species of American trees with their nearest European allies, all grown in close proximity and under as nearly as possible the same conditions. In the American species he finds, with the rarest exceptions, that the leaves fall earlier in the season, and assume before they fall a brighter tint; that they are less deeply toothed or serrated; that the buds are smaller; that the trees are more diffuse in growth and have fewer branchlets; and lastly, that the seeds are smaller—all in comparison with the corresponding European species." † The trees compared belong to several distinct orders, and are adapted to widely different stations, hence Darwin considers that the observed differences should be "attributed to the long continued action of a different climate."

More conclusive evidence of the direct effect of environment has been obtained in the case of Alpine plants. The especially characteristic features of these plants, as compared with similar or allied plants growing at lower levels, are a dwarfing in size and compactness of growth sometimes giving rise to a moss-like appearance; a more intense green colour in the leaves, and greater brilliancy and size in the flowers; an increased hairiness of the leaves, and occasionally a certain degree of fleshiness of the tissues. Now by growing lowland

* Proc. Acad. Nat. Sci. of Philadelphia, January 28, 1862.

† Quoted from "Animals and Plants," ii. p. 271.

plants at high altitudes, Bonnier,* Flahault,† and others have shown that such characters as these may be rapidly acquired. For instance, Bonnier made observations on *Teucrium Scorodonia* for no less than eight years, and he found that this plant, when sown at a high situation in the Pyrenees, produced very short aërial stems, with more hairy and darker green leaves, and more compact inflorescence. On the other hand, seeds gathered from plants growing at high altitudes, and sown in Paris, after three years produced elongated stems, with less hairy and brighter green leaves, or plants very similar to those from seeds obtained in the neighbourhood of Paris. The modifications acquired during a given time by a lowland plant grown at a high level, or a highland plant grown at a low level, took about the same amount of time to disappear, on returning the plants to their primitive climates. Again, Bonnier found that plants of *Lotus corniculatus* from Alpine situations had a very thick epidermis, a collenchymatous cortex, and a relative reduction of the wood. Those cultivated in lower altitudes had, on the other hand, a thinner epidermis, a cortex without collenchyma, and an increased development of wood.

With reference to the inflorescence, "Dr. Schûbeler sowed seeds of various plants in different latitudes in Norway, and proved that the brilliancy of the flowers increased with the latitude. So great were the differences that it was difficult to conceive that they were produced from the same batch of seeds. The differ-

* Ann. Sci. Nat. Bot., vii série, xx. p. 217, 1894.

† Ann. Sci. Nat. Bot., p. 159, 1879, and Rev. Gén. de Bot., ii. p. 513, 1891.

ences appeared in the first year. Similarly, seeds from Germany exhibited analogous differences." * Dr. Schûbeler also observed an increased greenness of the foliage.

The Arctic climate, though similar in many respects to the Alpine, yet differs considerably in others. By comparing plants from the Islands of Spitzbergen and Jan Mayen, with specimens of the same species collected in the Alps and the Pyrenees, Bonnier † has shown that there are modifications of structure corresponding to these differences of environment. The Arctic plants have more rounded cells and more considerable intercellular spaces in their tissues, whilst the cuticle is diminished in thickness, and the vessels are diminished in number and in calibre. These changes towards an incipiently aquatic type are probably due to the greater humidity of the air. The fleshiness of the leaves Bonnier attributes to the continuous solar illumination, though it may perhaps be due to the neighbourhood of the plants to the sea.

The effect of cultivation on the variation of plants is well known to be in many cases exceedingly great; but in hardly any of the recorded cases is any mention made of the extent to which artificial selection was practiced. One cannot tell, therefore, how much ought to be attributed to the direct action of the environment, and how much to selection. The following instance, however, seems to be the direct result of cultivation. It concerns the spiderwort, *Tradescantia virginica.*

* Quoted from Henslow's "Origin of Plant Structures," p. 118.

† Rev. Gen. Bot., vi. p. 505, 1894.

G. A. Brennan * records that he set out this plant in 1872, giving it very rich treatment. "In 1874 it began to deviate from the original trimerous type and to assume the tetramerous one, by developing another petal, and instead of doing this at the expense of the pistil or stamens, it *added* another sepal, another carpel with style, and two stamens, thus making a typically tetramerous flower." In 1876 a pentamerous plant was evolved; in 1879 a hexamerous; in 1882 a dimerous; and in 1884 a heptamerous. Thus as the result of thirteen years of cultivation, "a monocotyledonous plant has in bloom, at the same time, flowers of dimerous, trimerous, tetramerous, pentamerous, hexamerous, and heptamerous types respectively, each flower having twice as many stamens as sepals, petals, or carpels of ovary."

To turn to the Animal Kingdom, an interesting instance of variation following directly on change of environment is that noticed by Costa † in the oyster. Thus, on transferring young oysters from English shores to the Mediterranean, it was found that their manner of growth at once altered, and prominent diverging rays were formed, like those on the shells of the native Mediterranean oyster. The variations noticed by Leydig ‡ in the snail *Helix nemoralis* are attributed by him to the direct influence of environment, and this may be actually the case, but there is no evidence to prove it. He noticed that at Mainz the shell of this snail exhibits a fine citron yellow. This hue disappears further down the

* Amer. Naturalist, vol. xx. 551, 1886.
† Quoted from "Animals and Plants," ii. p. 270.
‡ Eimer's "Organic Evolution," p. 137.

Rhine, and at Bonn and in the still lower reaches the red of the snail deepens to a chocolate brown. Cockerell * also has noticed how sensitive is this species of snail to a changed environment. Thus it was introduced from Europe into Lexington, Virginia, a few years ago, and has since then varied extraordinarily. Already, in 1890, 125 varieties had been discovered in this locality. Of these no less than 67 were new, and unknown in Europe. The variations noticed by Gulick † in the land snails of the Sandwich Islands may also be due partly to the effects of environment. In a small forest region about forty miles by six miles in area, in the Island of Oahu, Gulick found about 175 different species, represented by 700 or 800 varieties. Successive valleys often showed allied species belonging to the same genus, and Gulick noticed that in every case, "the valleys that are nearest to each other furnish the most nearly allied forms; and a full set of the varieties of each species presents a minute gradation of forms between the more divergent types found in the more widely separated localities." Only a very few of the species ranged over the whole Island, most of them extending over only five or six miles, and a few over only one or two square miles. These variations did not *appear* to be due to the action of the environment, as the food, climate, and enemies in the different valleys seemed to be the same. Also the snails on the rainy side of the mountains did not differ any more from those on the dry side than they did from those inhabiting a neighbouring wet valley an equal distance away.

* Nature, vol. li. p. 393.

† Journ. Linn. Soc. (Zoöl)., vol. xi. p. 496.

As Wallace points out,* however, "it is an error to assume that what seem to us identical conditions are really identical to such small and delicate organisms as these land molluscs."

Upon Lepidoptera, we have seen in a previous chapter that the effect of particular conditions of environment, such as temperature and nutrition, may be considerable. One would imagine, therefore, that changes in the conditions of life as a whole might form an even more potent source of variation. Conclusive evidence upon this point is, unfortunately, almost unobtainable, though of the inconclusive kind which forms the larger part of this chapter there is plenty. For instance, Wallace † came to the conclusion that, with reference to the local forms occurring in the Indian and Malayan regions, "larger or smaller districts, or even single islands give a special character to the majority of their Papilionidæ. For instance: The species of the Indian region (Sumatra, Java, and Borneo) are almost invariably smaller than the allied species inhabiting Celebes and the Moluccas. The species of New Guinea and Australia are also, though in a less degree, smaller than the nearest species or varieties of the Moluccas. . . The species and varieties of Celebes possess a striking character in the form of the anterior wings, different from that of the allied species and varieties of all the surrounding islands. Tailed species of India or the Indian region become tailless as they spread eastward through the Archipelago; in Amboyna and Ceram the females of several species are dull-coloured, while

* "Darwinism," p. 148.

† "Contributions to Natural Selection," p. 167, 1870.

in the adjacent islands they are more brilliant." By actual measurement, Wallace found that "no less than fourteen Papilionidæ inhabiting Celebes and the Moluccas are from one-third to one-half greater in extent of wing than the allied species representing them in Java, Sumatra, and Borneo. Six species inhabiting Amboyna are larger than the closely allied forms of the northern Moluccas and New Guinea by about one-sixth. These include almost every case in which closely allied species can be compared." There are equally distinct local variations of form and colour. For instance, almost every Papilio inhabiting Celebes has wings of a peculiar shape, which distinguish it from the allied species of every other island. Thus the upper wings are more elongate and falcate, and the anterior margin is much more curved.

A remarkable instance of the direct effects of food, or perhaps of conditions of life in general, is mentioned by Darwin, who was himself informed of it by Moritz Wagner. "A number of pupæ were brought in 1870 to Switzerland from Texas of a species of *Saturnia* widely different from European species. In May, 1871, the moths developed out of the cocoons, and resembled entirely the Texan species. Their young were fed on leaves of *Juglans regia* (the Texan form feeding on *Juglans nigra*), and they changed into moths so different, not only in colour, but also in form, from their parents, that they were reckoned by entomologists as a distinct species." *

Reference has already been made in a previous chap-

* Quoted from Beddard's "Animal Colouration," p. 51.

ter to the observations made by Weldon * on the correlation between certain dimensions in local races of the shrimp. The degree of correlation was found by him to be practically constant, but the mean measurements themselves show distinct differences in the various local races. The variability or range of variation of the measurements about their mean shows much greater

LOCALITY.	MEAN TOTAL CARAPACE LENGTH.	PROB. ERROR.	MEAN POST-SPINOUS CARAPACE LENGTH.	PROB. ERROR.
1000 shrimps from Plymouth	249.63	4.55	177.53	3.50
800 " " Southport	248.31	3.96	180.29	3.65
500 " " Roscoff	251.51	3.32	178.00	3.02
380 " " Sheerness	247.33	3.29	179.68	2.91
300 " " Helder	251.38	4.36	181.67	4.02

differences still. As we see in the accompanying table, the total carapace length (expressed in terms of the body length, taken as 1000) varies in different localities from 247.33 to 251.38, or by 1.6 per cent. However, the probable error of variation of this dimension varies from 3.29 to 4.55, or by no less than 38.3 per cent. The mean post-spinous carapace length varies in its extreme limits by 2.06 per cent., and its probable error by 38.1 per cent. Now in that the shrimps differ so little in their average dimensions, they cannot be very divergent races, and hence one must conclude—supposing, of course, that the samples measured were fair ones, collected under similar conditions—that the differences in the variability of the shrimps obtained from the various regions are due chiefly to the

* Proc. Roy. Soc., vol. li. p 2, 1892.

action of a more or less correspondingly variable environment. It follows, therefore, that the environment at Plymouth and at Helder is more variable than that at Roscoff and at Sheerness.

To turn from marine Invertebrates to marine Vertebrates, the local races of the Mackerel have recently been studied in considerable detail by Garstang.* Some of the chief of the important results obtained by him are embodied in the accompanying table:

LOCALITY	NUMBER OF MACKEREL EXAMINED.	MEAN NUMBER OF				
		Transverse Bars.	Spotty Fish (per cent.).	Fin Rays in First Dorsal Fin.	Fish with 12 Fin-Rays in Second Dorsal Fin.	Fish with 5 Dorsal Finlets (per cent.)
Lowestoft	300	26.75	18	12.14 (Lowestoft and Ramsgate)	84.3 (Lowestoft and Ramsgate)	94.5 (Lowestoft and Ramsgate)
Ramsgate	100	26.88	28			
Plymouth	300	26.79	20	12.00	81.3	92.3
Scilly	74	26.82	18	12.16 (Scilly and Brest)	82.2 (Scilly and Brest)	93.0 (Scilly and Brest)
Brest	100	26.85	26			
Kinsale	410	27.15	19	12.14	85.4	94.4
Kerry	245	27.27	10	12.33	85.3	94.3
Newport, U. S. A.	100	27.38	66	11.88	63.0	97.0

The number of black transverse bars or stripes across the sides of the fish was found to vary from 23 to 33, 27 being in almost every instance the most frequently occurring number. The differences in the numbers of bars occurring in the various samples do not seem very great, but it is noticeable that all the samples from the North Sea and English Channel had invariably less than 27 bars (on an average), whilst those from the coasts of Ireland and America had more than 27. On

* Journ. Marine Biol. Soc., vol. v. p. 235, 1898.

classifying the fish according to the number of bars, it was found that, of those in the English Channel and North Sea (including Brest and Scilly), 20 to 22 per cent. had 28 or more bars; of those on the Irish coasts, 34 to 38 per cent.; and of those on the American coast, no less than 44 per cent. The proportions of fish having one or more round black dorso-lateral intermediate spots situated between the transverse bars, showed even more distinct differences. Thus 21 per cent. of the fish from the North Sea and English Channel were spotty; 22 per cent. of those from Brest and Scilly; only 15 per cent. of those from Ireland; but no less than 66 per cent. of those from America.

It was found that the number of fin-rays in the first dorsal fin varied somewhat according to the size of the fish, it being, for instance, 12.33 in Irish fish under 13 inches long, and 11.92 in those of 15 or more inches. To get rid of this variable factor, only fish 13 inches long were compared. Here again the American fish showed the greatest divergence from the general mean, whilst the Brest and Scilly fish were more or less midway between the North Sea and Channel fish on the one hand, and the Irish on the other. As regards the second dorsal fin, the variation in the number of fin-rays is much slighter than for the first dorsal fin, it being practically only from 11 to 13 (as against 10 to 15). The American fish showed a much wider variation than any of the others, only 63 per cent. of them having the modal number of 12 fin-rays, whilst the two samples of Irish fish showed least variation, 85.3 per cent. and 85.4 per cent. of them respectively having 12 fin-rays. In the number of dorsal finlets the American fish again showed the widest variation, only 79 per cent. of them having 5 finlets. The Irish fish again showed the smallest variation, 94.4 per cent. of them having 5 finlets, as against 93.6 per cent. in North Sea and Plymouth samples, and 93.0 per cent. in those from Brest and Scilly.

It is obvious, therefore, that the American mackerel constitute a distinct variety or race, the most noticeable characteristic of which is the high degree of spottiness. Garstang is of the opinion, also, that the mackerel which frequent the British coasts should be subdivided into two principal races: an Irish race, and an English Channel and North Sea race. The chief differences between these two subdivisions lie in the number of transverse bars and of spots, and to a lesser degree, of dorsal fin-rays and finlets. It is a striking fact, also, that "these peculiarities are greatest between the races of localities which are geographically remote, and least between those which occupy areas that are geographically contiguous. Between the mackerel of the North Sea and English Channel there are no differences at all; but the Irish race is distinctly divisible into two stocks, one of which is restricted to the west coast, and the other to the south."

Into the causes of the variations shown by these local races Garstang does not enter. It is highly improbable that all of the observed differences are the direct or indirect result of differences of environment, but it is possible that some of them, such as the bars and spots, and size of the fish (and with this the number of dorsal fin-rays), may be considerably influenced thereby.

The local races of the herring have been studied by Dr. Friedrich Heincke * with even greater minuteness than those of the mackerel by Garstang. Samples of herring from no less than a hundred different localities were examined, most of them in respect of about 25 different characters, and some in respect of over 50 characters. Heincke came to the conclusion that the various local races of herring examined by him differed from each other in the very characters which are used to differentiate the species of the genus Clupea from each other, though, as a rule, the differences were smaller. For instance, the most extreme variations noticed in the average number of vertebræ ranged from 57.6 in the Norwegian spring herring to 53.6 in the White Sea herring, or a difference of four vertebræ. The average number of vertebræ in the species Herring can be taken as 56, or eight more than in the Sprat, which can be taken as having 48. On the other hand the species *Clupea pilchardus* has, on an average, about 52 vertebræ, or does not differ any more from the sprat on the one hand, and the herring on the other, than do the most widely divergent local races of the herring.

Heincke found that, as a rule, the more widely the races are separated from each other geographically, or rather, the more their environmental conditions differ, the more do they differ in respect of certain characters. For instance, the number of vertebræ, and of scales between the ventral fin and the anus, showed the following mean variations:

* "Naturgeschichte des Herings," Abhandl. d. Deutsch. Seefischerei-Vereins, Bd. ii. Heft i. u. ii.

LOCAL RACE.			VERTEBRÆ.	SCALES.
Spring	herring,	Norway,	57.6	14.0
"	"	Schley,	55.5	13.7
"	"	Great Belt,	55.8	14.4
"	"	Rügen,	56.0	13.9
"	"	White Sea,	53.6	12.4
"	"	Zuidersee,	55.3	14.3
Autumn	"	E. Coast, Scotland,	56.5	14.8
"	"	North Sea (S. E.),	56.4	15.0
"	"	Jutland Bank,	56.6	14.5
"	"	Baltic Sea (W.),	55.7	14.5

Heincke seems to be of the opinion that these differences are largely the direct result of the environment, for he says that all the local races of herring are subjected to a very complex combination of environmental conditions, and that these react upon them to produce their especial characters. The White Sea herring is the most divergent from the general mean in respect of other characters besides the number of vertebræ and of scales. Thus it has only 2 to 10, or, on an average, 6 vomerine teeth, mostly in a single series, whilst other races have, on an average, 10 to 20 teeth, arranged in several series.

Heincke finds that the spring herring and the autumn herring are two more or less distinct races, not only in the Baltic Sea, but in other localities as well (West and East coasts of Scotland, North Sea, etc.). The spring herring differs from the autumn herring in that it is, as a rule, considerably larger; it has longer anal fins, and often a larger number of vertebræ. It always has a smaller number of keeled ventral scales, and a narrower skull, and very seldom has less than 9 rays in its ventral fins. Autumn herring with 8 ventral fin-rays occur fairly frequently, however (20 to 30 per cent. of all individuals). These differences of character Heincke attributes very largely to the different conditions of development. Thus, as regards the Western Baltic herring, the larvæ of the spring brood, developing in the warm and less saline waters of the Schley, reach the young herring stage within three or four months. Those of the autumn brood, on the other hand, which hatch in the more saline waters of the open sea, need the whole winter and spring, or 7 or 8 months, to reach the same stage.

The fish of the Atlantic and Pacific slopes have been studied and compared by Eigenmann.* In the eight families compared, the number of species on the Atlantic slope was more than twice as great as on the Pacific, but, nevertheless, the variation in the number of fin-rays in the Pacific species was greater in all but two of the families. The author considers that this may be due to the fauna being of diverse origin, and to its being comparatively new, and not yet in a state of equilibrium. The fish *Leuciscus balteatus* was studied in detail, and it was found that every locality in which it was examined had a variety peculiar to itself. As a rule, the lower the elevation of the locality from which the fish were obtained, the greater the number of fin-rays. The following are the mean values in support of this statement:

ELEVATION.	NUMBER OF SPECIMENS EXAMINED.	AVERAGE NUMBER OF RAYS.
1—750 feet, ...	189	18.4
1078—2000 feet, ...	234	16.6
2001—3100 feet, ...	388	17.5
5000 feet or more, ...	10	16.0

* Amer. Naturalist, xxix. p. 10, 1895.

From his extensive researches on the variation and distribution of mammals and birds in North America, J. A. Allen * has been able to arrive at several generalised conclusions concerning the geographical distribution of local races. Thus he finds that, as a rule, the mammals and birds of North America increase in size as we pass from the southern towards the northern regions. In the accompanying table are given the mean values obtained by him for the length of body,

SPECIES.	SEX	DIMENSIONS OF SOUTHERN SPECIMENS.			PER CENT. DIFFERENCE IN CORRESPONDING DIMENSIONS OF NORTHERN SPECIMENS.		
		Length.	Alar Extent.	Tail.	Length.	Alar Extent.	Tail.
Pipilo erythrophthalmus (towhee)	♂	7.88	9.88	3.56	+3.9	+14.6	−5.6
Agelæus phœniceus (red-winged blackbird)	♂	9.02	14.41	3.61	+1.6	+2.1	+ .6
Sturnella ludoviciana (meadow lark)	♂	9.81	15.70	2.85	+6.3	+3.8	+10.9
Sturnella ludoviciana (meadow lark)	♀	8.96	14.09	2.57	+6.6	+2;4	+9.7
Quiscalus purpureus (purple grackle)	♂	12.19	16.64	5.22	+3.6	+6.6	+1.5
Quiscalus purpureus (purple grackle)	♀	11.12	14.86	4.55	+3.0	+6.1	−1.3
Cyanura cristata (blue jay)	—	10.98	15.11	5.00	+6.6	+11.6	−2.2
Colaptes auratus (golden-winged woodpecker)	—	11.66	18.82	4.40	+6.8	+6.0	−1.1
Ortyx virginianus (common quail)	♂	9.46	14.16	2.52	+7.6	+9.0	+11.9
Ortyx virginianus (common quail)	♀	9.37	14.02	2.54	+4.9	+7.7	+5.1

stretch of wings, and length of tail of seven different species of birds. In the middle portion of the table are given the actual values (in inches) for the Southern races, from Florida, whilst in the right half are given

* Bull. Mus. Comp. Zoöl. Harvard, vol. ii. p. 161–490; *also* Cope's "Factors of Organic Evolution," p. 45.

the percentage variations, on these values, of the corresponding values for the Northern races (from Northern States, Massachusetts, and Southern New England). On an average fifteen specimens were measured in each case, the extreme numbers varying between 6 and 40. As regards body length, we see that the Northern forms invariably exceeded the Southern, the average difference amounting to 5.1 per cent. In alar extent they were likewise invariably greater, the average excess being 7.0 per cent. In tail measurement, however, the difference was not nearly so constant, it being greater in the Southern races than in the Northern in four out of the ten sets of measurements, whilst the average excess amounted to only 2.9 per cent.

Accompanying the increase in size of the Northern forms, Allen finds that, as a rule, there is an appreciable decrease in colour. In the South, dark-coloured birds, such as the red-winged blackbird, become blacker. The slaty and olive tints of other birds, and the various shades of red and yellow, become far more intense as one proceeds south, and the pigmentation of the bill and feet also increases. Allen says "the difference in colour between the extremely Northern and extremely Southern representatives of a given species is often so great that, taken in connection with other differences, as in the general size and the size and form of the bill, the two extremes might excusably be taken for distinct species." The size of the bill varies, as a rule, in the inverse ratio to the size of the body, and "in many species there is not only a marked *relative* increase in the size of the bill to the southward, but in some an *absolute* increase, especially in its length." This increase is quite marked in the genera *Quiscalus*, *Agelæus*, *Geothlypis*, *Troglodytes*, *Seiurus*, etc.

As to the causes of these geographical variations, it is of course impossible to ascribe them with any certainty even to the indirect effects of change of environment, much less to the direct. Still, as Allen points out, there is often a somewhat close correlation between geographical varieties and the meteorological peculiarities of the regions in which they occur, which suggests a connection of some sort between the two. The increase in colour towards the south coincides with the increase in the intensity of the sun's rays, and in the humidity of the climate. The increase in colour observed in birds on passing from East to West seems also to coincide with an increase of humidity, "the darker representatives of any species occurring where the annual rainfall is greatest, and the palest where it is least." This coincidence occurs not only in the birds of the United States, to such a degree that Allen says he knows of no exception, but in Europe also. Thus birds from the Scandinavian coast are very much darker than in central Europe, where the rainfall is only half as great. Allen says that this correlation of brighter and deeper tint with increased humidity is exhibited by the mammals of these districts, as well as by the birds.

The differences in the local races of certain Mammals are even more striking than in those of the birds. The Canidæ, for instance, are represented in North America by six species, viz., gray wolf, common fox, gray fox, coyote, arctic fox, and kit fox, of which the first three are the widest ranging species. Allen found that, in respect of skull measurement, the common wolf is fully a fifth larger in the northern parts of British America and Alaska than it is in northern Mexico, the southern limit of its habitat, whilst, as we see in the accompanying table, specimens from intermediate regions show a gradual intergradation between these extremes.* The common fox from Alaska is about 10 per cent. larger than that in New England,

SPECIES.	LOCALITY.	NUMBER OF SPECIMENS MEASURED.	MEAN LENGTH OF SKULL.	MEAN WIDTH OF SKULL.
Gray wolf	Forts Simpson, Yukon and Rae,	9	10.38 in.	5.40 in.
	Forts Benton and Union,	16	9.45	5.07
	Forts Kearney and Harker,	9	9.69	5.18
	Rio Grande and Sonora,	3	8.37	4.31
Common fox	Alaska,	9	5.98	3.20
	Mackenzie River District,	18	5.80	3.02
	Upper Missouri,	9	5.78	2.90
	Essex County, New York,	12	5.40	2.80

whilst the gray fox probably varies considerably more in size with locality, but the number of skulls obtained for measurement (15 in all) is insufficient to warrant any generalisation in its case.

This increase in size on passing from South to North is not universal, however. Thus lynxes and wild cats, though series of skulls were obtained from such widely separated localities as Alaska, California, and Northern Mexico, revealed no appreciable variation of size with

* U. S. Geol. and Geographical Survey, vol. ii. p. 309, 1876.

locality. Panthers and ocelots, indeed, showed a very considerable increase in size on passing southward. Still the increase in size of Carnivorous Mammals on passing from South to North may be taken to be a very general rule. In addition to the Felidæ mentioned, this relation is well shown in the badger, marten, wolverine, and ermine.

Of other Mammals, the relationship between locality and size is well shown by members of the deer family, the Virginia deer affording an especially striking instance. The Glires (squirrels, marmots, mice) also increase, as a rule, towards the North. For instance, the northern race of flying squirrels is half as large again as the southern, but these two extremes are connected by a complete chain of intermediate forms. As in the case of birds, mere size of body is not the only characteristic which varies with locality. The ears and the feet may undergo considerable changes in addition. Thus in mammals with large ears, such as wolves, foxes, some of the deer, and especially the hares, there is often a striking increase in the size of these appendages on passing from North to South. The ears of the little wood hare (*Lepus sylvaticus*), found in Western Arizona, are nearly twice the size they attain in the variety found in more easterly and northerly regions. Again, in *Lepus callotis* the ear is one-third to one-fourth larger in examples obtained from Mexico than in those from Wyoming, whilst the little brown hare (*L. trowbridgei*) shows a similar increase in the size of the ear in the south.

Darwin * has collected several cases in which climate

* "Animals and Plants," ii. p. 268.

had an influence on the hairy covering of animals. Thus he says "Dr. Falconer states that the Thibet mastiff and goat, when brought down from the Himalaya to Kashmir, lose their fine wool. At Angora not only goats, but shepherd-dogs and cats, have fine fleecy hair, and Mr. Ainsworth attributes the thickness of the fleece to the severe winters, and its silky lustre to the hot summers. Burnes states positively that the Karakool sheep lose their peculiar black curled fleeces when removed into any other country."

What may be termed, perhaps, the classical instance of the formation of a local race through changed conditions of life, is that of the Porto Santo rabbit.* A female rabbit and her young were turned loose on the Island in 1418, and they increased so rapidly as to become a nuisance, and finally caused the abandonment of the settlement. The present-day form of these rabbits shows very considerable differences from the original form. Thus the two measured by Darwin were only 14½ and 15 inches in length, instead of the 17 or 18 inches of the English rabbit. The weight of one of them—which had, however, become somewhat thin from living in captivity—was only 1 pound 9 ounces, four English wild rabbits averaging 3 pounds 5 ounces. The limb bones were smaller in the proportion of rather less than five to nine. In colour, the Porto Santo rabbits have a redder upper surface, rarely interspersed with any black or black-tipped hairs, and in none of the seven specimens examined by Darwin had the upper surface of the tail and the tips of the ears any of the blackish gray fur which

* "Animals and Plants," i. p. 118.

is generally regarded as one of the specific characters of the rabbit. Finally, two male Porto Santo rabbits, when kept in captivity, never lost their extreme wildness, and would never associate or breed with the females of various breeds placed with them.

To what extent these remarkable changes were the direct result of a changed environment, it is, of course, impossible to say; but it was proved that at least the colouring was a direct effect. Thus one of the feral rabbits, after being kept for four years in captivity, was found by Darwin to have acquired a blackish gray fur on the upper surface of the tail and the edges of the ears, whilst the whole body was much less red; *i. e.*, it had recovered the proper colour of its fur after four years of English climate.

The influence of domestication combined with artificial selection is well known to everyone, but what shares of the changes produced are to be assigned to each of these agencies is, as a rule, quite indeterminable. However, one may with some reservation ascribe to domestication changes effected in characters which have never been the subject of selection. For instance, the weight of the rabbit was found by Darwin to increase on domestication, the result, probably, both of more ample feeding and of artificial selection. The skull capacity, on the other hand, by no means proportionately increased in size, and as this is scarcely a character on which selection would be practised, we may consider the relative diminution as probably due to the direct influence of domestication. The reason why we cannot say with absolute certainty that it is a direct effect, lies in the fact that the character of "skull capacity" may be more or less closely correlated with some other character which has been the object of selection, and so have been thereby unintentionally modified. In the accompanying table are given the mean values of Darwin's measurements:*

BREED OF RABBIT.	WEIGHT.	LENGTH OF BODY IN INCHES.	LENGTH OF SKULL IN INCHES.	CAPACITY OF SKULL.	RATIO OF BRAIN CAPACITY TO LENGTH.
7 various wild rabbits . .	2 lb. 15 oz.	17.1	3.09	950	100
3 Porto Santo rabbits . .	(1 lb. 9 oz.)	(14.75)	2.88	828	101.1
4 various domestic rabbits . .	3 lb. 4 oz.	19.75	3.47	864	78.8
7 large lop-eared rabbits . .	7 lb. 4 oz.	24.62	4.11	1136	83.1

The weight of most breeds of domestic rabbit is not much greater than that of wild ones, but that of the lop-eared variety is more than twice as great. The length of body was measured from incisors to anus, whilst the capacity of the skull was determined by weighing the small shot taken to fill it (the numbers given in the table being the weight in grains). Taking the relation of capacity of skull to length of body in the wild rabbit as 100, we see that, on an average, the skull capacity of the domestic rabbit is about 20 per cent. less. That of the Porto Santo rabbit is, on the other hand, very slightly greater.

The diminution in the size of the rabbit's brain is attributed by Darwin to the effects of disuse, and

* "Animals and Plants," i. p. 133.

he ascribes certain of the changes in other domesticated animals to a similar cause. Thus he found the length of the sternum in eleven different breeds of domestic pigeon to be on an average 13.0 per cent. shorter than in the wild rock pigeon.* The crest of sternum, scapulæ, and furculum were also reduced in size, but the wings were slightly increased, owing to the greater length of the wing feathers. Again, in eight out of the eleven breeds of fowl examined, the weight of humerus and ulna, relative to that of femur and tibia, was, on an average, 24.2 per cent. less than in the wild *Gallus bankiva*, and in all eleven breeds the depth of the crest of the sternum (to which the pectoral muscles are attached), relative to its length, was diminished, the average diminution being 17.5 per cent.† In the case of the duck, Darwin weighed the entire skeleton, as well as individual parts, and he found that whilst in the four breeds of domestic duck examined the weight of the femur, tibia, and tarsus, relative to that of the body, was, on an average, 28.5 per cent. *greater* than in the wild mallard, that of the humerus, radius, and metacarpus was 9.0 per cent. *less*.‡ That this decrease in the weight of the wing bones is the direct result of disuse was proved by the fact that in a domestic call duck which was in the habit of constantly flying about for miles, the relative weight of the wing bones was actually 12.1 per cent. greater than in the wild mallard.

* "Animals and Plants," i. p. 184.

† *Ibid.* i. p. 285.

‡ *Ibid.*, i. p. 301.

PART III.

VARIATION IN ITS RELATION TO EVOLUTION.

CHAPTER XI.

THE ACTION OF NATURAL SELECTION ON VARIATIONS.

"*Selection does nothing without variability, and this depends in some manner on the action of the surrounding circumstances on the organism*" (*Darwin*, "*Animals and Plants*," i. p. 7).

"*The foundation of the Darwinian theory is the variability of species*" (*Wallace*, "*Darwinism*," *p.* 41).

"*What forms the basis of the constant 'individual variations' which, after the precedent of Darwin and Wallace, we regard as the foundation of all processes of natural selection?*" (*Weismann*, "*Germ-Plasm*," *p.* 410.)

Proof of Natural Selection in the crab and in the sparrow—Selection in man—Evolution of the mouse—Inheritance of acquired characters seems to be shown by cumulative effects of conditions of life, as European climate acting on American maize; domestication acting on wild turkeys and ducks; changed climate acting on sheep and dogs—Environment may act on germ-plasm through specific excretions and secretions—Cases of inherited effects of use and disuse, and of epilepsy, accounted for—Somatic variations may increase variability, and so afford Natural Selection a better handle to work upon.

THE contents of this chapter are well summarised in the three quotations given at its head. It deals with variations in their relation to Natural Selection, and with the gradual evolution thereby brought about. The fundamental importance of variations in the evolutionary process has been dwelt on again and again by Darwin, by Wallace, and by most of the subsequent writers on the subject, and as this doctrine is so universally admitted, it is unnecessary to discuss it any further here. At the present day, however, there appears to be a considerable amount of scepticism among some men of science as to the extreme importance which has been generally attached to the agency of Natural Selection. Some, such as Driesch, have even denied its existence altogether, whilst many others hold that its existence has never been demonstrated. They hold with Lord Salisbury * that "no man, as far as we know, has ever seen it at work." The evidence to be adduced will show, I believe, that this statement is erroneous, but even if it be correct, it cannot disprove the theory, the validity of which seems to me a logical necessity. Thus, granted the geometrical rate of increase possessed by all organisms, and the severe struggle for existence thereby entailed; granted that all organisms show individual variations, and, to a considerable extent, hereditary transmission of such variations, then it must follow that, on an average, more of the organisms possessing favourable variations better adapted to their environment will survive than of those possessing less favourable ones. That is to say, the species will become gradually modified by the action of Natural Selection.

Numerical evidence in support of the theory of Natural Selection has been obtained only quite

* "Presidential Address, British Association," 1894.

recently, and this is not to be wondered at, considering the numerous and extended observations it, as a rule, entails. In the case of the small shore crab, *Carcinus mænas*, however, Professor Weldon * has succeeded in overcoming most of the inevitable difficulties and pitfalls, and has obtained evidence which, though at present not absolutely convincing, yet has a very high degree of probability of truth. In 1893 Mr. H. Thompson carefully determined the relation of the mean frontal breadth to the carapace length in male crabs collected at a particular patch of beach in Plymouth Sound. The mean breadth was found to vary very rapidly with the length of the crab, hence its value was determined separately in small groups of crabs, differing from each other by not more than .2 mm. Twenty-five such groups, for crabs between 10 and 15 mm. long, were measured in respect of frontal breadth. A similar series of measurements was carried out by Thompson on crabs collected at the same spot in 1895, and another by Weldon on crabs collected in 1898. On comparing the three series of data thus obtained, it was evident that the mean breadth of crabs of a given carapace length had *steadily decreased*. For instance, in crabs with a carapace length of 11.5 mm., the frontal breadth had a percentage length of 79.72 in 1893, 78.88 in 1895, and 78.40 in 1898. Again, in 14 mm. crabs, it had a length of 76.26 in 1893, 75.44 in 1895, and 74.44 in 1898.

It would seem, therefore, that the frontal breadth of these crabs is diminishing, year by year, at a very rapid rate. This Professor Weldon attributes to a selective

* Report of Brit. Assn., 1898, p. 887.

destruction, caused by certain rapidly changing conditions in Plymouth Sound. Owing to the building of a huge breakwater, the scour of the tide has been diminished, and the large quantities of china clay carried down by the rivers from Dartmoor into the Sound therefore settle in increasing quantities in the Sound itself. Also the quantity of sewage and refuse finding its way into the Sound is steadily increasing, owing to the increase in the size of the contiguous towns and dockyards. "It is well known," says Professor Weldon, "that these changes in the physical condition of the Sound have been accompanied by the disappearance of animals which used to live in it, but which are now found only outside the area affected by the breakwater." In order to test his supposition of selective destruction, Professor Weldon placed a number of crabs in a large vessel of sea water, in which a considerable quantity of very fine china clay was suspended. The clay was prevented from settling by a slowly moving automatic agitator. After a time, the dead crabs were separated from the living, and both were measured. In the figure given below is shown the result obtained.

Here the upper curve shows the distribution of frontal breadths of the 248 male crabs experimented on, and the dotted curve the distribution of frontal breadths of the 94 survivors. The line 0 represents the mean frontal breadth of all the crabs, the dotted line S the mean of the survivors, and the dotted line D the mean of the dead crabs. The crabs which survived thus had a distinctly smaller frontal breadth than those which were killed, just as the 1898 crabs had a smaller breadth than the 1895 ones, and these than the 1893 ones. There seems no reason to believe that the action of the mud upon the beach is different from that in an experimental aquarium, and hence, in Professor Weldon's opinion, there is "no escape from the conclusion that we have here a case of Natural Selection acting with great rapidity because of the rapidity with which the conditions of life are changing." The selective destruction seems to depend on the filtration of the water into the gill chambers of the crabs. To quote Professor Weldon, "The gills of a crab which has died during an experiment with china clay are covered with fine white mud, which is not found in the gills of the

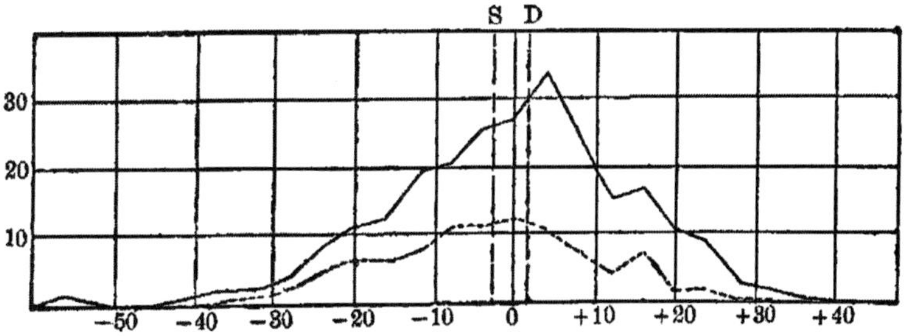

FIG. 26.—Distribution of the frontal breadths of 248 male crabs, and of the 94 survivors.

survivors. In at least 90 per cent. of the cases this difference is very striking." Professor Weldon thinks it can be shown that a narrow frontal breadth renders one part of the process of filtration of water more efficient than it is in crabs of greater frontal breadth.

Such, then, is Professor Weldon's demonstration of a particular instance of Natural Selection. In order to strengthen the proof of its existence, further measurements of crabs collected at the same spot a few years hence ought to be made, as Professor Weldon himself well recognises, in order to see whether the destructive process is still continuing. If this is the case, and if crabs measured in, say, 1903 and 1908 show a further diminution of frontal breadth, then the evidence in favour of selection would amount to a very high degree of probability indeed. Owing to the changing relation of its parts with growth, the crab is a somewhat unsatisfactory organism to work with, and, indeed, the apparent change observed between 1893 and 1898 might possibly, though not probably, owe its origin to quite another cause than Selection. For instance, the conditions of environment such as temperature, nutrition, and purity of the water may have acted directly on the crabs so as to retard their growth. Now Professor Weldon assumes that all crabs of, say, 12 mm. length are approximately the same age, but obviously this need not be so from year to year. Under less favourable conditions, the moulting may have gone on as usual, but the rate of growth have been reduced. Now we have seen that the frontal breadth diminishes very rapidly with growth, and hence it might happen that the narrower fronted 12 mm. crabs of 1898 are narrower simply because they are *older* than were the more favourably situated 12 mm. crabs of 1893. Mr. J. T. Cunningham * has pointed out that in 1893 the temperature of the Channel waters was abnormally high, and he considers that this produced a more rapid growth of the crabs, and hence, for a given size of crab, an apparent increase of frontal breadth. However, Professor Weldon † does not believe that the tempera-

* Nature, vol. lviii. p. 593.

† *Ibid.*, p. 595.

ture of the beach where his crabs were collected, in that it looks due south and is uncovered for hours daily, was any lower in 1898 than in 1893, and also he found that crabs gathered in January were no narrower fronted than those gathered in August, as they ought to have been on Cunningham's hypothesis.

The proof of the existence of Natural Selection really centres upon the proof of a selective destruction or death rate. If among any group of organisms the eliminated individuals can be measured and examined, as well as the survivors, and if it be found that these two divisions differ in their mean characters, then Natural Selection *must* have been at work. Very likely the parts or organs measured do not represent the characters upon which the selective process had been acting, but are merely correlated with them. But that is no matter. The offspring of the survivors will have different average qualities from those of the previous unselected generation, or the race will become modified by Natural Selection.

Unfortunately in the majority of cases, as in Professor Weldon's crabs, it is impossible to get hold of the eliminated individuals, and hence the proof of Natural Selection is rendered much more laborious, and at the same time more open to possible source of error. In the case of the (introduced) English sparrow (*Passer domesticus*), however, Bumpus * has succeeded in obtaining the desired material. One hundred and thirty-six of these sparrows were collected after a very severe storm of snow, rain, and sleet in North America, and of these 72 revived, whilst 64

* Biol. Lectures, Wood's Holl, 1898, p. 211.

perished. On comparing the survivors with the eliminated individuals, very appreciable differences were found to exhibit themselves. The means of the values obtained with all the birds, both male and female, are given in the accompanying table:

	MEAN VALUES.			ARITHMETIC MEAN ERROR.		
	SURVIVORS.	ELIMINATED.	PER CENT. DIFFERENCE.	SURVIVORS.	ELIMINATED.	PER CENT. DIFFERENCE.
Total length,	158 mm.	160 mm.	+1.27	2.51	3.48	+38.6
Alar extent,	245 mm.	245 mm.	±0.0	4.20	4.60	+ 9.5
Weight,	25.2 gm.	25.8 gm.	+2.38	10.9	12.6	+15.6
Length of beak & head	31.6 mm.	31.5 mm.	— .32	5.51	5.64	+ 2.4
Length of humerus,	.736 inch	.728 inch	−1.09	.016	.0201	+25.6
Length of femur,	.716 "	.709 "	— .98	.014	.020	+42.9
Length of tibio-tarsus	1.138 "	1.128 "	— .88	.0294	.0338	+15.0
Width of skull,	.603 "	.601 "	— .33	.010	.012	+20.0
Length of sternum,	.845 "	.834 "	—1.30	.032	.038	— 3.1

Here we see that the average characters differ but little. The eliminated individuals are 1.27 per cent. greater in length, and 2.38 per cent. greater in weight, whilst they are about 1 per cent. smaller than the survivors in respect of most of the other characters measured. The variability, or range of variation of the eliminated birds about their mean, is, however, very much greater than in the case of the survivors. Of the nine characters measured, the variability is greater in eight, the average excess being no less than 18.8 per cent. The variability was less in respect of the sternum alone, and then only by 3.1 per cent. In the accompanying figure are given curves of distribution of the lengths of the surviving and of the eliminated birds. It is obvious that the dotted line curve, which represents the eliminated individuals, is, on the whole, much more flat-topped than the other curve. The very long individuals seem especially handicapped in the struggle for existence, as of the 18 birds obtained in which the length was 164 mm. and upwards, no less than 14 perished. Also the two shortest birds obtained perished. The conclusion which Bumpus draws from these most interesting observations is as follows: "Natural Selection is most destructive of those birds which have departed most from the ideal type, and its activity raises the general standard of excellence by favouring those birds which approach the structural ideal." The observations really show more than this, however. It is

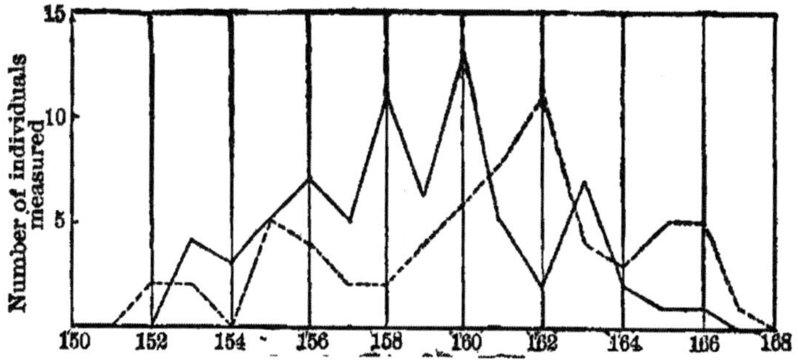

FIG. 27.—Distribution of the lengths of surviving and of eliminated sparrows.

true that, as a rule, the most extreme individuals in either direction are eliminated, but if the distributions of the various characters be plotted out as in the above figure, it will be seen that in the case of some of the other characters, as in that of length, the eliminating process acts much more on the extreme individuals in one direction than on those in the other.

For instance, the accompanying figure shows the distribution of the weight values of the birds. The curves are very irregular, but it is obvious that the dotted line curve is shifted distinctly to the right, indicating that the eliminated birds were, on an average, heavier. This conclusion has already been obtained by the simple process of taking averages; but the curves show in addition that it is the *very* heavy birds which were more

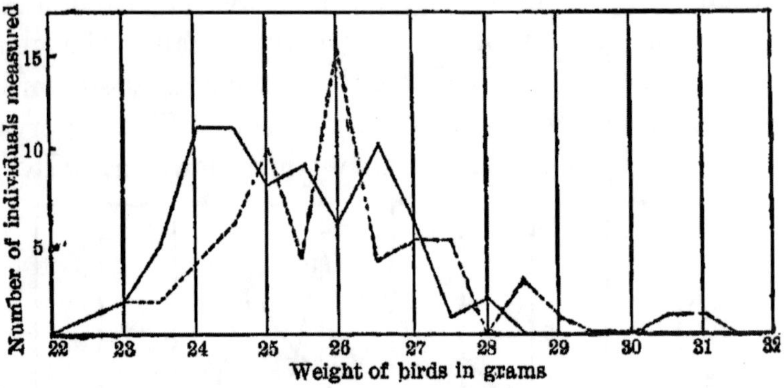

Fig. 28.—Distribution of the weights of surviving and of eliminated sparrows.

especially eliminated. Thus of the 14 birds of 27.3 gms. and upwards obtained, only three survived. Similarly also in respect of the femur measurements, it was found that of the 19 birds obtained with a femur length of .685 inch or less, only 7 survived whilst 12 were eliminated.

The next generation of birds collected in the storm-swept area would accordingly be shorter in length, weigh less, have longer legs, have a longer sternum and a greater brain capacity than the former generation; supposing, of course, that the variations existing in these characters were partly of blastogenic, and not wholly of somatogenic origin; and this could scarcely fail to be the case. Several of the changes in characters, especially of the total length, weight, and femur length, might possibly be present, on an average, to just as marked an extent as in their parents (the survivors of the previous generation); for though the characters would tend to undergo reduction by virtue of their "regression towards mediocrity," yet they would tend to be enhanced by reason of the fact that more of the extreme individuals (which would be of proportionally greater weight in determining the characters of the next generation) had been weeded out than of the mediocre ones. Bumpus does not give any details as to the way in which the sparrows were collected, and whether the sample obtained was representative of all the sparrows in the area in question. Supposing it were not, and it included only sparrows which were exposed to the force of the storm through failing to get adequate shelter, then, of course, the average change produced in the characters of the next generation would be much less than that suggested by the above figures.

Professor Weldon * has adopted a very ingenious method for determining the presence or absence of Natural Selection in the case of a certain terrestrial mollusc, *Clausilia laminata*. The outer layer of the shell in this and other pulmonates is secreted by the growing edge of the mantle once and for all, and it undergoes practically no subsequent change. The upper whorls of an adult shell therefore afford an unaltered record

* Biometrika, i. p. 109, 1901

of the condition of the young shell, from which this adult was formed by the subsequent deposition of new material. By measuring the upper whorls of the adult shells, one is accordingly able to determine the characters possessed, not by *all* young shells, but by the young shells which were successful in attaining the adult condition. How would measurements on such adult shells compare with those on young and growing shells, some of which would almost certainly undergo destruction before attaining their full development? To answer this question, Professor Weldon measured with great exactness the radius of the spiral at various (angular) distances from the apex of the shell in 100 adult individuals, and also in 100 young individuals of less than half their length. The means of the values so obtained were practically identical in the two classes of shells, so it seems to follow that the mean spiral of young shells is not altered during growth by any process of selective destruction. On the other hand, the variability of the radial spiral measurements was considerably greater in the young shells than in the adult ones (on an average, in the proportion of 120 to 100 for the first whorl and a half.) Hence we may conclude that during the growth of this mollusc some processes are at work which effectually eliminate the abnormal shells more rapidly than the others, and so diminish the variability of the survivors. As the average character of the race does not undergo any change, it follows that the abnormalities eliminated are evenly distributed about the mean. Such a process of selection has been termed by Professor Pearson * *periodic.*

* "Grammar of Science," p. 413.

The proof given by Professor Pearson * of the existence of a selective death rate in the case of man seems to me scarcely to entitle him to speak of it as a case of "Natural Selection." Thus he shows that there is a fairly close correlation ($r = .26$) between the ages at death of brother and brother, and a less close one between those of fathers and sons ($r = .12$ to $.14$). For instance, the mean age at death of men not dying as minors is 61 years. If, however, one brother of a pair dies at 25, then the other will, on an average, die at 51.6, or 9.4 years earlier than the mean; if one dies at 85, then the other will, on an average, die at 67.2, or 6.2 years later. There is something in the constitution of a man, therefore, which to a large extent determines when he shall die, or undergo elimination. His death is not at all a matter of chance. Further analyses of data by Miss Beeton and Professor Pearson † indicate the same thing, though they also lead to other and somewhat unexpected conclusions. Thus, from the pedigree records of members of the society of Friends, the authors found that older (adult) brothers and sisters on an average live distinctly longer than younger (adult) brothers and sisters, and that the greater the difference in age, the greater the difference in expectation of life. For instance, a man born 6 years after his elder brother will probably live 4 years less than he will; one born 10 years after, 7 years less, and one born 17 years after, as much as 12 years less. Put in other words, the eldest children of a family have the best chance of life,

* "Grammar of Science," p. 497; *also* Beeton and Pearson, Proc. Roy. Soc., lxv. p. 290, 1899.

† Biometrika, i. p. 50, 1901.

and the youngest the worst. We may perhaps look upon this decrease of vitality as the direct result of the diminished vigour of the parents at the time of conception, and of the mother during the period of intrauterine development. If this is actually the case, then we ought to find that the expectation of life is more closely correlated with the age of the mother at the time of conception than with that of the father.

Again Miss Beeton, in conjunction with Mr. G. U. Yule and Professor Pearson,* have found that there is a direct correlation between the duration of life in parents, and the number of children borne by them. It was found that fertility was correlated with longevity even in parents of 50 years and upwards, when the fecund period is passed, though the correlation is not by any means so close as in parents under 50.

For instance, American mothers dying at 25 had on an average 2.2 children; those at 35, 4 children; and those at 50, 5.7 children: but mothers dying at 70 had on an average 6.8 children, and those at 90, no less than 7.6 children. With English mothers dying at 50 years and over, the increased fertility is not so marked, and it becomes slightly diminished in those living over 75 years. Similarly also with English fathers the relation of fertility to longevity is less marked than in the case of American fathers.

All these data may undoubtedly be taken to indicate, therefore, that longevity is inherited in man, and as long life means a healthier and stronger constitution, it is natural to find that it also betokens increased

* Proc. Roy. Soc., lxvii. p. 159, 1900.

power of procreating offspring. It does not necessarily follow, however, that Natural Selection, in the ordinary sense of the term, is at work. The time of death may be quite uncorrelated with any particular structural characters of the body, but be dependent only on the so-called vigour or vitality of the organism. Each subsequent generation may therefore be more "vigorous" than the one before it, owing to the elimination of a portion of the less vigorous individuals, but as, in all probability, there is always a tendency to the production in each generation of a certain number of unfit individuals, or a slight diminution in the average vitality of the whole group, it would follow that a certain amount of elimination is always necessary, to enable a race to maintain its average vitality from one generation to the next. Certainly, in the case of the human race, there is no evidence that the average vigour and vitality are increasing. Everything goes to prove rather that they are on the wane.

Man is therefore an unsatisfactory organism in which to determine either the existence or the non-existence of Natural Selection. His conditions of death are as unnatural as his conditions of life. The usual cause of his death, disease, counts for little or nothing amongst the lower animals, whilst the usual causes of death amongst them, namely, want of food and natural enemies, count for little or nothing with man. To prove the existence of *Natural* Selection, one must choose for observation an organism living under natural conditions.

A very interesting case of the formation of a local race through the probable agency of Natural Selection has recently been described by H. L. Jameson.* On the north side of Dublin Bay there is a tract of sand-hills, running along the coast for about three miles. It is separated from the mainland by a tidal channel about a quarter of a mile wide at high water, but only 20 yards or so at low water. These sandhills are thickly populated with mice, which were noticed by Jameson to harmonise strikingly in colour with the sand. Traps were set, and altogether 36 mice were caught. The specimens varied considerably in the shade of their fur, showing every gradation from the typical *Mus musculus* of the farmhouses in Ireland and England to individuals with extremely pale dorsal fur—usually of a rufous or fulvous gray colour—pale buff ventral surface, and pale and fulvous appearance of the hairs on the ears, tail, and other parts of the body. Also the feet were white or pale buff, instead of the smoky gray or white of the ordinary House-mouse, whilst the claws were flesh-coloured. Of the 36 specimens, only five were of more or less the typical colour, four were slightly paler, and the remaining 27 markedly pallescent.

These mice differ in other characters also. Thus, if the adult specimens be split up into three groups, according to their colouration, and means taken of the measurements made by Jameson, the values given in the table below are obtained. Though the number of measurements is so small, there can be little doubt that the tail of the palest individuals is distinctly longer than that of the typical ones. Perhaps also the head and body and the hind foot are

* Journ. Linn. Soc. (Zoöl), vol. xxvi. p. 465, 1898.

slightly shorter, though one cannot speak with any certainty. The habits of the mice had changed in addition, as they were found to burrow their own holes, no holes burrowed by other animals being available, as in the case of the typical wild mouse.

The development of the protective colouration and habits probably owes its origin to the short-eared owls and hawks which were noticed to frequent the sandhills, and which would more readily perceive and capture the darker mice. These would gradually be weeded out, therefore, whilst the light-coloured individuals would survive and propagate their more favourable characteristics.

NUMBER OF SPECIMENS MEASURED.	COLOURATION.	LENGTH IN MILLIMETRES OF		
		Head and Body.	Tail.	Hind Foot.
7	Typical or slightly pale	81.3	76.5	18.1
9	Markedly pale	76.3	77.8	16.9
20	Still more pale	80.2	80.0	17.2

Perhaps the most interesting point of all about these observations is that it has been found possible to fix a time limit for the duration of the evolutionary process. The sandbanks are known to be gradually increasing in area, and, by a careful study of old maps, Jameson found that previous to 1780 they did not exist at all. In 1810 the island was only a quarter of a mile long, so we may conclude that the pale race of mice has had not more than about a hundred years for its evolution.

Are Acquired Characters Inherited? We see that

Evolution is brought about by the action of Natural Selection on variations, it selecting some and rejecting others, and so gradually altering the average characters of the race; but are blastogenic or germinal variations alone of value to such a selective agency, and are somatic variations, or so-called acquired characters, valueless in this respect? As is well known, the question of the heritableness of acquired characters has been one of the most hotly debated of all biological problems, and is one which even now separates most biologists into two opposite and apparently irreconcilable camps. It behoves us, therefore, to see if we cannot find some *via media*, which, though unable to admit of the heritableness of localised tissue changes such as injuries and mutilations, is yet able to adopt reasonable evidence, both experimental and theoretical, in favour of a partial inheritance of certain generalised tissue changes.

The chief experimental evidence in support of the apparent heritableness of acquired characters lies in the numerous and undoubted proofs of the cumulative action of conditions of life. Of such proofs, one of the most striking is that recorded by Darwin * with reference to the effects of a European climate on the American varieties of maize. Thus Metzger cultivated in Germany a tall kind of maize, *Zea altissima*, brought from the warmer parts of America, and, "During the first year the plants were twelve feet high, and a few seeds were perfected. . . In the second generation the plants were from nine to ten feet in height, and ripened their seed better. . . Some of the seeds had

* "Animals and Plants," i. p. 340.

even become yellow, and in their now rounded form they approached common European maize. In the third generation nearly all resemblance to the original and very distinct American parent-form was lost. In the sixth generation this maize perfectly resembled a European variety."

Other instances of the cumulative effects of conditions of life on plants have already been recorded in former chapters. Thus Lesage found that if Garden cress were treated with salted water, a much more marked effect was produced in the second year than in the first, the alteration effected in the tissues of the second generation seeming to be carried on from the point gained in the first. Bonnier found that seeds of *Teucrium scorodonia* gathered from plants growing at high altitudes, and sown in Paris, only produced plants showing nearly similar characters to the local forms after three years' exposure to the new environment.

Among members of the Animal Kingdom the evidence is no less conclusive. Thus Darwin * records that "Dr. Bachman states that he has seen turkeys raised from the eggs of the wild species lose their metallic tints and become spotted with white in the third generation." Again, Mr. Hewitt, who often reared ducks from the eggs of the wild bird, and who was careful that there should be no crossing with domestic breeds, "found that he could not breed these wild ducks true for more than five or six generations, as they proved so much less beautiful. The white collar round the neck of the mallard became much broader and more irregular, and white feathers ap-

Ibid., ii. p. 250.

peared in the ducklings' wings. They increased also in size of body; their legs became less fine, and they lost their elegant carriage. Fresh eggs were then procured from wild birds, but the same result followed." Again, Darwin* records that "according to Pallas, and more recently according to Erman, the fat-tailed Kirghisian sheep, when bred for a few generations in Russia, degenerate, and the mass of fat dwindles away, the scanty and bitter herbage of the steppes seems so essential to their development." The fleece of sheep imported from Europe to the West Indies is much affected, and "after the third generation, the wool disappears from the whole body, except over the loins; and the animal then appears like a goat with a dirty door-mat on its back. A similar change is said to take place on the West Coast of Africa." Another instance of the effect of climate on sheep is recorded by Brewer.† Sheep taken from southeastern Ohio, a district noted for its excellent wool, and pastured on the alkaline soil of a certain portion of Texas, had the texture of their wool much altered, and its reaction to dyes showed obvious differences. Brewer states that "the change in the character of the wool begins immediately, but is more marked in the succeeding fleeces than in the first. It is also alleged that the harshness increases with succeeding generations, and that the flocks which have inhabited such regions several generations produce naturally a harsher wool than did their ancestors, or do the newcomers."

The deteriorating effect of an Indian climate on

* *Ibid.*, i. p. 102.

† *Vide* Cope's "Factors of Organic Evolution," p. 435.

dogs is well known. Hounds seem to degenerate most rapidly of all, whilst greyhounds and pointers also decline rapidly. Darwin * was informed by Dr. Falconer that bull-dogs "not only fall off after two or three generations in pluck and ferocity, but lose the underhung character of their lower jaws; their muzzles become finer and their bodies lighter." He also mentions a case of a pair of setters, born in India, which perfectly resembled their Scotch parents. Several litters were raised from them in Delhi, but none of the young dogs obtained resembled their parents in size or make, their nostrils being more contracted, their noses more pointed, their limbs more slender, and their size inferior. On the coast of Guinea, "dogs, according to Bosman, alter strangely; their ears grow long and stiff like those of foxes, to which colour they also incline, so that in three or four years they degenerate into very ugly creatures; and in three or four broods their barking turns into a howl."

Darwin considers this tendency to rapid deterioration in European dogs may be largely attributed to reversion. It is of course possible that this may be the case, but it seems to me more probable that it is due to the direct and cumulative effects of changed conditions of life.

The cumulative effect of conditions of life is admitted, even by Weismann, in the case of the butterfly *Polyommatus phlæas*. As already mentioned in Chapter VII., this occurs as a reddish gold variety in Germany and other countries of similar latitude, and as a much darker variety in more southerly countries, as

* *Ibid.*, i. p. 39.

Italy and Greece. Though these forms can be more or less transformed into each other, by suitable exposure of the pupæ to warmth or cold, yet Weismann found that from German pupæ he could never obtain butterflies so dark as the darkest forms of the southern variety, whilst from Neapolitan pupæ he could never get them so light as the ordinary German variety.* It seemed to him, therefore, "that the two varieties may have originated owing to a gradual cumulative influence of the climate, the slight effects of one summer or winter having been transmitted and added to from generation to generation." Weismann explains this case of apparent transmission of acquired characters by supposing that the temperature slightly affects the determinants of the wing scales contained in the germ-plasm, as well as more markedly influencing the determinants of the rudimentary wings in the chrysalis. Moreover he suggests that "in many other animals and plants influences of temperature and environment may very possibly produce permanent hereditary variations in a similar manner."

This suggestion of Weismann's contains in it, it seems to me, the germ of an idea which further observation and experiment may prove to be of fundamental importance in evolution. The idea itself is no new one, and has probably occurred independently to many writers. As far as I am aware, it was first suggested by Galton,† when propounding the theory of heredity to which that of Weismann bears so striking a resem-

* "Germ-Plasm," p. 399.

† Proc. Roy. Soc., xx. p. 394, 1872; Contemporary Review, December, 1875; Journ. Anthropol. Inst., 1875, p. 346.

blance. Thus he concluded that we are almost justified in reserving our belief that the body cells can react on the sexual elements, *i. e.*, that acquired characters can be inherited; but he himself proposed to accept the supposition of their being faintly heritable. More recently, Cope* has embodied the idea in his "Theory of Diplogenesis." Thus he says, "Now, since these somatic cells develop the modifications which constitute evolution in their subsequent growth into organs, there is no reason why the reproductive cells which experienced similar influences should not develop similar characters, so soon as they also are prepared to grow into organs. . . The effects of use and disuse are twofold, viz.: the effect on the soma, and the effect on the germ-plasm. . . The character must be potentially acquired by the germ-plasma, as well as actually by the soma." However, when Cope begins to briefly expand his theory, he seems to me to drift into improbable and unverifiable speculations. Thus he imagines that the transmission of external influences is primarily through the nervous system—perhaps through the organisation of some peculiar mode of motion—and secondarily through nutrition.

In order to account for the numerous instances of the cumulative effects of changed conditions of life, it seems, therefore, that we may assume with considerable probability and reason that the germ-plasm is directly affected as well as the body tissues. These apparent instances of the inheritance of acquired characters are in reality, therefore, nothing of the kind, but are due to the germ-plasm reacting to change of en-

* Amer. Naturalist, xxiii. p. 1058, 1889.

vironment simultaneously with the body tissues. As Weismann points out, a necessary corollary to this view is "the assumption of material determinants which exist in the germ-plasm and are passed on from one generation to another." If change of environment acts cumulatively on the fleece of the sheep, or the structural characters of a dog, it follows that it must in each of the first few generations act also on the "determinants" in the germ-plasm specifically representing such specific characters. The effect produced on such determinants in the first generation must serve more or less as a starting point for the environment to work upon still further in the next generation, and so on.

Through what agency is the environment enabled to act on the germ-plasm? To me the only conceivable one is a chemical influence, through products of metabolism and specific internal secretions. We have seen in a previous chapter that the products of metabolism of an organism may exert a retarding effect on its own growth, and in some cases a stimulating effect on the growth of other organisms. Physiological research of the last few years has shown that most of the organs and tissues of the body have specific internal secretions, which, passing into the general circulation, may exert an influence of vital importance on the general metabolism of the organism. Thus extirpation of the thyroid gland produces symptoms which in many animals end fatally, but which may be diminished or suppressed by feeding on the gland substance, or injection of extracts of it. Extirpation of the suprarenal glands results in much more speedy death, and here again the injection of extracts may delay the fatal issue. Extirpation of the pancreas causes the production of severe diabetes, and ultimately death, but such an effect may be avoided by the grafting of a portion of excised gland in the peritoneal cavity or the tissues. In such a case it cannot, of course, exercise its digestive function, but its internal secretion prevents the onset of the fatal diabetes. Again, extirpation of the total kidney substance of a dog leads, not to a diminished secretion of urine, but to a largely increased secretion, accompanied by a rapid wasting away which soon ends fatally. Hence the kidneys may possess an influence on the metabolism of the whole body, as well as their obvious secretory function. The spleen appears to have an internal secretion which is of influence in setting free the pancreatic ferment. Finally, extracts of various nervous tissues, brain, spinal cord, and sciatic nerve, have been found when intravenously injected to produce a distinct fall of blood pressure, whilst those of the pituitary body produce a marked rise.

Does it not seem distinctly probable, therefore, that every tissue in the body to some extent affects every other tissue? Each may have its own specific products of metabolism, and perhaps specific internal secretions, which, passing into the general circulation, may in turn stimulate or depress, or otherwise affect, every other tissue in the body. Whenever a changed environment acts upon the organism, therefore, it to some extent affects the normal excretions and secretions of some or all of the various tissues, and these react not only on the tissues themselves, but also to a lesser degree upon the "determinants" representing them in the germ-plasm.

It should be mentioned that the influence of somatic variations on the germ-plasm through the agency of various secretions has already been suggested by Delage.* Though he does not admit Weismann's doctrine of determinants, he thinks that the ovum may contain specific substances of an identical nature to those contained in the cells of the principal classes of tissues, such as the nervous, muscular, and perhaps glandular. Conditions of life such as climate and food, which through the intermediation of the blood influence the constituents of certain of the body tissue cells, will therefore influence the same substance in the ovum, or produce hereditary variations.

The hypothesis of specific secretions is of distinct help in accounting for certain apparent instances of the inherited effects of use and disuse. As we have seen in a former chapter, Darwin found that the relative size of the brain of the domestic rabbit has considerably diminished. Possibly this may have been the result of more ample food, and of artificial selection of individuals with large bodies and small heads, and of panmixia (cessation of Natural Selection), but it seems almost more probable that it is due, at least in part, to the inherited effects of disuse. Thus a rabbit, when kept in captivity, would need to use its brain but little, and hence the excretions and secretions of the nervous tissues would be diminished. The "determinants" in the germ-plasm corresponding to these would be less stimulated than in wild rabbits, and

* "Heredité," pp. 806 to 812, Paris, 1895.

hence in the next generation the development of the brain (and probably the other nervous tissues) would take place somewhat less vigorously, and the adult brain be in consequence somewhat diminished in size. In the next generation the diminution would be greater still, and so on.

Again, we have seen that in man, for instance, the degree of (hereditary) pigmentation of the skin seems to vary closely with the intensity of the heat and light experienced. It is possible that the specific excretory products of the pigment deposited in the skin, as a direct response to the action of the environment, may stimulate the pigment "determinants" in the germ-plasm to increased vigour, so that in the next generation the organism will tend to become slightly more pigmented than it had been in the previous one. Supposing, on the other hand, the pigment cells of the skin received no light rays whatsoever, as in animals which had wandered into a subterranean cave, their metabolism would be reduced almost to nil, and so the pigment "determinants" in the germ-plasm would diminish in vigour, and the offspring of the animals would be (at birth) somewhat less pigmented than they had been in previous generations.

It is obvious that on our specific secretion hypothesis only a certain class of acquired characters can be in any degree heritable; only those, in fact, of which the corresponding tissues possess a specific secretion or excretion, capable of acting specifically on the "determinants" of such tissues in the germ-plasm. For instance, the blacksmith cannot transmit his brawny arm in any degree to his descendants, as it is scarcely possible that the arm muscles can have a secretion different from that of the other muscles of the body. The greater muscular development of the man as a whole, however, may lead to the production of slightly more muscular children than the average.

On our hypothesis, the heritableness of mutilations and injuries is not admissible. It is almost inconceivable that each spot of skin on the body, or each finger, should have a specific secretion, and that an injury to it, by changing its secretion, should so affect the germ-plasm as to produce a similar change in the corresponding area of skin or the finger of the offspring. How, then, is it possible to account for the various apparent instances of inherited injuries, such as are quoted by Eimer,* Cope,† and others who believe in the transmissibility of such characters? There certainly seem to be a small number of thoroughly well authenticated cases, but the number is so small that we may perhaps attribute them to mere coincidence. The millions of instances of injuries which show no trace of any transmission provoke no remark, as it is only what we are led by common experience to expect. Supposing, on the other hand, a child exhibits any birth mark or deformity bearing some similarity to an injury or mutilation in a parent, it is at once hailed as a remarkable case of inheritance of an acquired character.

There are, however, certain cases of the apparent inheritance of acquired characters which require more detailed criticism. These are the well-known experiments and observations of Brown-Séquard on injuries

* "Organic Evolution," p. 173.
† "Factors of Organic Evolution," p. 431.

of the nervous system in guinea-pigs. As Brown-Séquard experimented over a period of thirty years on thousands of guinea-pigs, it might be thought that we could accept his results as absolutely conclusive. Yet a repetition of some of his experiments by Romanes and by Hill seems to show that they may be very largely erroneous. Thus, like Brown-Séquard, Romanes * found that some of the progeny of parents in which an injury to the restiform body had produced protrusion of the eyeball, showed a protrusion likewise, though this was less marked, and always affected both eyes; but it seemed that this might be an accidental occurrence, in that normal guinea-pigs are sometimes to a certain extent exophthalmic. Again, Romanes found that some of the progeny of animals in which hæmatoma and dry gangrene of the ears had supervened after injuring the restiform body, also became affected. However, the morbid state seemed to arise at any time in the life history of the individual, and the process not only affected a much less quantity of the ear, but also a different part of it. One therefore might imagine it to be due to mere coincidence, or to transmitted microbes; but Romanes does not think this can be the case, as, on the one hand, he has never seen the peculiar morbid process of the ears in other guinea-pigs, and, on the other hand, he was unable to inoculate the ears of healthy animals with matter from the ears of mutilated guinea-pigs.

Romanes repeated Brown-Séquard's experiments on the section of the cervical sympathetic nerve, but he never observed in their progeny any change in the shape

* "Darwin and after Darwin," vol. ii. p. 104 *et seq.*

of the ear or partial closure of the eyelids. Dr. Leonard Hill* has also repeated them with some thoroughness. The operation was performed on six guinea-pigs, and these animals were allowed to interbreed. It was again performed on twelve of their offspring, and these were also allowed to interbreed, but none of the young of either the first or the second generation showed any persistent droop of the eyelid. Hill found, however, that many of the young guinea-pigs exhibited a partial closure of the eye for some time after birth, but this phenomenon was due entirely to conjunctivitis, the result of dirt getting into the eyes. It affected both eyes equally often, and when it terminated the droop disappeared also. One is strongly tempted to conclude that the partial closure of the eyelids observed by Brown-Séquard was due to a similar cause, and was no more hereditary than in Hill's guinea-pigs. Certain of Brown-Séquard's experiments have, however, been corroborated by subsequent observers, and must therefore be accepted. Thus he found that animals which had been rendered epileptic by injury to the spinal cord, or section of the sciatic nerve, might transmit this epilepsy to their offspring. These results have been confirmed by Obersteiner,† and Westphal has even succeeded in producing epilepsy, which was transmitted to the offspring, by striking guinea-pigs on the head with a hammer. It has been suggested by Weismann that the transmission might be due to the introduction of some microbe into the operative wound, which both caused epilepsy in the parent, and, by invading the germ cells,

* Proc. Zool. Soc. 1896, p. 785.

† Oesterreichische medicinische Jahrbücher, 1875, p. 179.

produced it in the offspring also. This could not have been the case in Westphal's experiments, however, as in them no wound at all was made.

How, then, can this apparent transmission of acquired characters be accounted for? Our hypothesis of internal secretions supplies a very simple explanation. Thus the secretions from the brain of an epileptic guinea-pig, no matter how this epilepsy had been produced, would almost certainly be abnormal. Even supposing that they were without effect on the "determinants" of the nervous tissues in the germ-plasm, it is a very probable supposition that they might so affect the growth of the nervous tissues of the offspring, during intra-uterine development, as to provoke a similar abnormal condition in them.

In mammals and other viviparous animals, it is probable that changed conditions of life produce part of their cumulative action during the period in which the embryo is under the influence of the maternal fluids. It is of course possible that *all* of the cumulative effect is then produced, though in such a case we should have to find some other explanation than that given above of the cumulative effects noticed in oviparous animals as *Polyommatus phlœas*. In the case of the gradual degeneration of the pure bred dog under an Indian climate, for instance, the environment may so act upon the maternal parent as to produce slight changes in the body tissues, and also to alter the character of the secretions and excretions. These, acting on the offspring during their embryonic development—when, as we have seen in a previous chapter, the tissues are extraordinarily sensitive to their environment

—may produce more obvious degenerative changes, which will of course continue and be increased during extra-uterine growth. This second generation of dogs, besides being modified in external characters, will therefore have the nature of their internal secretions more altered than had the first generation. These changes will react still further on *their* offspring during intra-uterine development, and so on.

Our conclusions as to the reaction of the germ-plasm to the external conditions of environment place a much higher value on somatic variations as a factor in Evolution than that accepted by Weismann and his followers. It is for this reason that the effects of environment in the production of variations have been dealt with at such length in the preceding chapters. Every obvious effect produced in an organism by the direct action of the environment, *may*, in my opinion, be accompanied by a more or less corresponding, though much slighter, effect upon the determinants in the germ-plasm, and express itself in the next generation as an apparently cumulative effect of the changed environment. How often this possible influence on the germ-plasm actually shows itself, and what may be the numerical measure of its extent, can only be determined by long continued observation and experiment.

As we shall see in the next chapter, somatic variations may be of very great importance in evolution by reason of their adaptiveness to sudden changes of environment; but, quite apart from any question of adaptation, it is probable that they may be of value in affording Natural Selection a better chance of exerting its influence. How this is so, is best explained by means of a diagram.

FIG. 29.—Effect of variable environment on variability of organisms.

The upper of the two accompanying figures represents a (roughly) normal curve of distribution of 324 measurements (represented by dots and crosses). As is indicated on the base line, these vary in size from the general mean by from ±1 to ±10 per cent. Supposing that the larger individuals (the crosses) were better adapted to their environment, and were being gradually selected by the agency of Natural Selection, whilst the shorter ones were gradually being weeded out, then it is probable that the selective process would act only very slowly, as the range of variation is so slight. Thus differences of 2 or 4 per cent. from the average are so small as to be almost inappreciable, whilst greater differences, as of 7 or 8 per cent., are exhibited by only a very small proportion of the whole (only 12 out of the 324 measurements being 7 per cent. greater than the average, and only 6 of them 8 per cent. greater). Now let us suppose that this group of 324 individuals is acted on by a variable environment, so that the slight range of blastogenic variations is enhanced by the superposition of somatic modifications. Out of every ten organisms of any particular size, let one be increased by 2 per cent., another by 4 per cent., another by 6 per cent., another by 8 per cent., and another by 10 per cent., owing to the action of a favourable environment, whilst the other five of the ten organisms are diminished by similar amounts, owing to the action of an unfavourable environment. The lower figure shows the new distribution of the organisms according to their altered magnitudes. For instance, of the 40 crosses representing individuals 1 per cent. larger than the average, 4 are increased by 2 per cent., and so are placed in the + 3 per cent. column, and 4 are diminished by 2 per cent., and so are placed in the − 1 per cent. column. Four more are placed in the + 5 per cent. column, and 4 in the − 3 per cent. column, and so on. A similar process was applied to all the other measurements, as far as possible, the 6 individuals 7 per cent. larger than the average being, for instance, increased and diminished by ± 4, 6, and 8 per cent. The curve of distribution of the 324 individuals now takes the form of the lower figure. We see that it is much more flat-topped, indicating that the range of variation is much greater than before. This is, in fact, more than doubled, the arithmetical mean error being increased from ± 3.2 per cent. to ± 6.8 per cent. The individuals now vary in size by ± 17 per cent., so Natural Selection can act with much greater celerity and certainty than before. Thus no less than 32 of the 324 individuals are now 11 per cent. or more larger than the average, and so offer a very appreciable handle for Selection to work upon. The distribution of the dots and crosses shows us, also, that all the extremely large individuals are also individuals which were larger than the normal before the variable environment was brought to bear on them. Many of the larger individuals were rendered smaller than the average by the action of an unfavourable environment, and many of the smaller rendered larger by a favourable environment—*i. e.*, there has been a good deal of mixing of the individuals as originally distributed—but the fact remains that the *extremely large* individuals, which Natural Selection would be especially likely to favour, and the *extremely small* ones, which it would be especially likely to eliminate, are still those which were originally, as the result of blastogenic variation, respectively larger and smaller than the average. The selected individuals are therefore not only larger in themselves, but in that their "largeness" is to some extent a blastogenic variation, their offspring will also be, on an average, larger than the normal.

It is not intended to imply that increased variability is by any means always an advantage. In a stable form, upon which Selection is acting but little, it might be a distinct disadvantage, as the more variable individuals might be less adapted to their environment than the less variable. In the case of the sparrow, for instance, we saw that Bumpus found that the extreme individuals in either direction tended to be weeded out, though there was a much greater elimination of the extreme individuals in one direction than of those in the other. What is true for one form, however, is by no means necessarily true for another, and in a rapidly evolving organism, such as the pale-coloured mouse found on the sand-banks off Dublin, it is probable that the eliminated individuals would be chiefly confined to the darkest specimens, and include but few of the palest ones.

CHAPTER XII.

ADAPTIVE VARIATIONS.

Adaptability a fundamental property of protoplasm—Instances of adaptive variation in plants—Acclimatisation of Protozoa to high temperature, to poisons, to mechanical stimuli, to saline solutions —Acclimatisation of fresh-water Mollusca to salt water, and of various marine animals to fresh water—Acclimatisation of Mammals to vegetable poisons, and to toxins—Sum total of somatic variations always in direction of adaptation—Somatic variations of importance in evolution, but they can effect little without Natural Selection—Germinal Selection.

THE question of the definiteness or indefiniteness of variations has been frequently and hotly debated, but there has been a singular absence of exact definitions of the views actually held by the supporters of the rival theories. Had such definitions been forthcoming, I doubt if any fundamental differences of opinion would have been found to exist at all. Take, for instance, Darwin's definition of definite variations, viz.: "The effects of (conditions of life). . . may be considered as definite when all or nearly all the offspring of individuals, exposed to certain conditions during several generations, are modified in the same manner." * There is surely nothing in this definition which would not be generally admitted. As has been shown at some length in several of the preceding chapters, change in one or many of the conditions

* "Origin of Species," p. 6.

of life may lead to very considerable changes in the form and structure of all or most of the organisms exposed, even in one generation. Hence, if Darwin's definition be accepted as it stands, we are compelled to admit that variations may be definite. Supposing, however, it be taken to imply the cumulative, and so hereditary, action of conditions of life acting for several generations, then those who refuse to admit the validity of the instances of such cumulative action adduced in the last chapter might also refuse to admit the existence of definite variations. But assuming the former interpretation as the correct one, are we to agree with Henslow * that "in nature variations are *always* definite," or are we to follow Darwin in believing that variations are, as a rule, indefinite, and only exceptionally definite? Here, it seems to me, we are in want of more exact definitions. Probably Henslow would admit that the variations in the number of carpels in the common daisy, or of veins in the leaf of the beech tree, or of stigmatic bands on the seed capsules of the poppy, are governed by the laws of chance, or if he did not, how could he account for the fact that the frequencies of distribution of the respective numbers are in accordance with the Law of Error? Clearly, in such cases the variations must be indefinite. Supposing, however, that the distribution of the variations in the length of the leaves of a plant grown upon land occur according to the laws of chance, whilst that of the leaves of the same species of plant grown in water also follows these laws, but supposing also, that the average length of the aquatic plant leaves is considerably

*" Original of Plant Structures," p. ix.

greater than that of the land plant leaves, then obviously we should have here a case of both definite variation and indefinite variation. The leaves of the aquatic plant would have varied in the direction of greater length, or would have varied *definitely* in adaptation to their new environment, but the distribution of their variations about their mean would still be in accordance with the laws of chance, or would be *indefinite.*

The term "definite," as applied to variations, seems to be generally regarded as more or less synonymous with "adaptive." Thus Lloyd Morgan * defines definite or determinate variations as "variations along special or particular lines of adaptation," while Henslow † says "Definite variations are always in the direction of adaptation to the environment itself." Hence it seems to me that the discussion of the definiteness or indefiniteness of variations may, for practical purposes, be narrowed down to the following questions: (1) Have conditions of life an appreciable influence on organisms, and if so, (2) Is this influence in any case cumulative, *i. e.*, partly inherited, and (3) How far are the effects produced adaptive? The first two questions I have already endeavoured to answer in the preceding chapters. The third we will now proceed briefly to inquire into.

As far as the limited number of observations available can show, *adaptability would seem to be a fundamental property of protoplasm.* Whenever an organism is exposed to changed conditions of life, then it is found that the original want of adaptation becomes gradually and progressively diminished with increase

* "Habit and Instinct," p. 311.

† *Ibid.*, p. viii.

in the duration of the exposure. In most animals, the change in the direction of adaptation is slight, but it is probably always there, if only it be carefully looked for. In plants it is, as a rule, greater, and may be obvious to the most cursory observation. Instances of it have already been described at some length in preceding chapters, and hence it is unnecessary to do more than briefly recall these here. We saw that Karsten found that a kidney bean reared in the dark for a month or two weighed 20 per cent. more than one reared in the light, yet, owing to the absence of the stimulus of light, its leaves did not weigh a fifth as much. Lothelier found that plants such as *Berberis vulgaris* bore non-spinescent leaves in a moist atmosphere, but spines and spines alone in a perfectly dry one. Costantin found that he could change the form of *Hippuris* at will, by growing the aquatic form of the plant on land, and the terrestrial form in water. All the leaves produced under water were long, undulated, and delicate, whilst those in air were short, erect, and firm. Costantin, and also Godron, obtained very similar results by growing other aquatic plants on land, and terrestrial ones in water, the change being always in the direction of adaptation to the new surroundings. Again, Lesage found that by watering various plants with water containing salt they developed characters similar to those exhibited by maritime plants, viz., increased thickness of leaves, larger and more numerous palisade cells, and diminution of the intercellular spaces and of the chlorophyll. Bonnier found that plants of *Teucrium scorodonia*, when grown at a high situation in the Pyrenees, exhibited features characteristic of alpine plants, viz.: very short aërial stems, with hairy and dark green leaves, and compact inflorescence. Seeds gathered from these plants and sown in Paris after three years produced elongated stems, with less hairy and brighter green leaves, or plants very similar to those from seeds obtained in the neighbourhood of Paris.

In addition to changes of climate and soil, plants can adapt themselves also to mechanical stresses and strains. Thus Ray * sowed a mould (*Sterigmatocystis*) in two vessels, one of which was fixed, and the other subjected for two months to a rapid oscillatory movement. Instead of a thick feltwork of mycelium, this latter vessel contained small, perfectly spherical, elastic masses consisting of entangled filaments. The supporting tissues of the plant were strengthened in response to the violent mechanical strains, the membranes being twice or three times as thick, and the filaments having many more partition walls. Again, R. Hegler † found that "the hypocotyl of a seedling sunflower, which would have been ruptured by a weight of 160 gm., bore a weight of 250 gm. after having been subjected for two days to a strain of a weight of 150 gm. The weight was subsequently increased to 400 gm. without injury. . . Leaf stalks of *Helleborus niger*, which broke with a weight of 400 gm., were able to resist one of 35 kgm. after having been subjected to a strain for about five days." Thus protoplasm has the power of responding to, and counteracting the action

* C. R. Acad. Sci., cxxiii. p. 907.

† Ber. Verhandl. K. Sachs. Gesell. Wiss., v. p. 638, 1892 (quoted from Henslow's "Origin of Plant Structures," p. 204).

of, external mechanical forces by the formation of supportive tissues. It is by reason of this power that plants grow vertically upwards in opposition to the force of gravity.

Some of the most remarkable instances of adaptation in plants are those relating to the interchange between roots and stems. In many cases, at least, it would seem that when a subterranean root becomes aërial, its characters tend to approach to those of a stem, whilst a normally aërial stem, grown underground, develops the characters of a root. For example, "An old acacia with a decaying trunk sent down an aërial root from the living part, about six feet from the ground. When it had been rooted in the soil for some time, it became detached by the wind; the root then became a 'stem,' the upper part putting out foliage." * Again, Dr. Lindley records that "a young willow tree had its crown bent down to the ground; this was covered with earth, and soon emitted an abundance of roots. The true roots were then removed from the soil, and the stem inverted. The roots now became branches and emitted an abundance of buds, and the tree ever afterwards grew upside down." Accompanying such changes of function are found corresponding changes of histological structure. Costantin † determined the effect of growing stems of brambles underground, and he found that the number and volume of the cortical cells increased, the collenchyma disappeared, the liber fibres diminished or disappeared, and starch could be formed and stored up in the parenchymatous tissues.

* Quoted from Henslow, *ibid.*, p. 179.

† Bull. de la Soc. Bot. de Fr., p. 230, 1883.

Moreover these modifications were uniform, affected all the tissues, and were rapidly produced, a week or two sufficing.

There is no evidence, as far as I am aware, to show how widespread is this phenomenon of interchange between roots and stems, and hence one cannot accept it as a generalised property of plants. In any case one must bear in mind that it may not, after all, be a case of direct adaptation to surroundings in the ordinary acceptation of the term, but may be the calling up, in response to one of two stimuli, of one of two groups of characters long since acquired by the plant protoplasm.

A case of adaptation which appeared to be to some extent hereditary has recently been recorded by Errera.* Conidia of the mould *Aspergillus niger* were cultivated by Dr. Hunger for two generations in Raulin's nutritive solution, to which 6 per cent. of common salt had been added, and when placed in a similar salt solution they were found to produce spores in 3¾ days. Conidia which had been cultivated in the salt Raulin solution for only one generation took 4 days to produce spores, however, whilst those which had been cultivated in Raulin solution containing no additional salt took 5 days. On the other hand, when some of the conidia cultivated under the three sets of conditions were placed in ordinary Raulin solution, those kept two generations in salt solution showed only slight sporification in 5 days, those kept one generation showed more marked sporification, whilst those kept throughout in ordinary Raulin solution spored in 4 days. Spores from these three last cultures in normal Raulin

* Bull. Acad. Roy. Beligque, p. 81, 1899.

solution were then sowed in a solution to which 18.4 per cent. of salt had been added. After 5 days the original normal Raulin culture showed no germination; that originally kept one generation in salt Raulin solution showed slight germination, and that originally kept two generations distinct germination. Thus the adaptation to a concentrated salt solution was not entirely lost even after rearing in a normal medium, or was in some degree inherited, especially in the case of the greater degree of adaptation produced by the growth of two generations in salt solution. Doubtless this "inheritance of acquired characters" was due to the salt solution influencing the germ cells at the same time as the body cells. The same explanation may be used to account for the somewhat similar results obtained by Ray * with *Sterigmatocystis alba.* Conidia of this mould were sown in a solution of dextrose, the development taking place but slowly, owing to the want of adaptation to the new environment. On continuing the culture in the sugar solution, however, the rate of development gradually increased from generation to generation, till, finally, the sixth generation showed a more abundant development after 8 days than the first one had after 15 days. The morphological characters were progressively modified in addition, so that the mould came finally to resemble a *penicillium.*

In the members of the Animal Kingdom, the power of adaptation is, as a rule, far less marked than in those of the Vegetable Kingdom, but probably it is present to a greater or less extent in all organisms, from the lowest to the highest. In certain Flagellata, for in-

* Rev. Gen. de Bot., ix. p. 193, 1897.

stance, Dallinger* has demonstrated a most remarkable and extreme adaptability to high temperature. Starting at 15.6° C., he gradually raised the temperature of the water containing these monads up to 70.0° C., when the experiment was ended by an accident. The exact times in the course of the experiment at which a given temperature was reached are not mentioned, but from the description afforded they are gathered to be roughly those given in the accompanying table:

Original temperature,	15.6° C.	After 36 months,	34.4°
After 4 months,	21.1	" 39 "	38.9
" 6 "	22.8	" 41 "	41.7
" 8 "	23.3	" 48 "	58.3
" 15 "	25.6	" 60 "	58.9
" 23 "	26.7	" 61 "	61.1
" 32 "	33.9	(Several months more)	70.0

From this we see that the experiment, as far as it was carried, lasted about six years. The raising of the temperature was not by any means even, the organisms frequently reaching stages at which for months at a time an increase of temperature of half a degree or less was immediately followed by adverse effects, and in some instances by the death of many of the organisms. For instance, when a temperature of 25.6° had been reached, it was found that for a space of five months the temperature could not be raised even .3 of a degree without a distinctly evil effect being produced. In fact, it was found that the progress of acclimatisation at lower temperatures was, as a rule, much slower than at high ones. Thus, within a space of seven months, it

* Journ. Roy. Microsc. Soc., vol. vii. p. 191, 1887.

was found possible to raise the temperature of the organisms from 41.7° to 58.3°. Also to raise the temperature from 61.1° to 70.0° took only a few months (number not stated). As to the absolute upper limit of temperature these infusoria can withstand, it is of course impossible to judge, but there seems no reason to suppose that it might not be considerably higher than that reached by Dallinger. A striking proof of the altered condition of the organisms was furnished by the fact that some of those acclimatised to 70.0°, died off when placed in a suitable nutritive solution at 15.6°.

This acclimatisation was probably for the most part a direct adaptation of the protoplasm to its new environment, but it must also have been in part due to natural selection. Dallinger noticed on more than one occasion that a good many of the organisms were killed off, and these would doubtless have been the less adaptable ones, the more adaptable surviving. Still, as far as one can judge from the brief account given, the temperature was often raised over considerable intervals without any such fatalities.

Dallinger's results, in addition to their intrinsic value, are of great interest in that they enable us to account for the presence of various Protophyta, such as *Oscillatoriæ* and *Nostocaceæ* in hot springs. The temperature of many of these springs is considerably above 60.0° C., and that of the California geysers, in which *Nostocaceæ* (possibly *Protococcus*) are found, reaches the remarkable temperature of 93°. Certain metazoa, also, are stated to live at temperatures considerably above 45°, or temperatures which prove fatal to their allies. All such instances as these * are probably due to gradual acclimatisation, accompanied by a variable amount of selective destruction of the less adaptable organisms.

Other observations on acclimatisation among Protozoa have been made by Davenport and Neal.† The acclimatisation of *Stentor cœruleus* to weak corrosive sublimate solution was tested. Stentors kept for two days in .00005 per cent. solution were found, on immersion in .001 per cent. solution of sublimate, to be killed off after (on an average) 304 seconds' exposure. Stentors kept in pure water, on the other hand, were killed after only 83 seconds' immersion. Similar results were obtained in other experiments, it appearing that within certain limits the resistance period varied directly with the strength of the solution in which the protoplasm had been cultivated. If, however, the culture solution were too strong, (above .0001 per cent.), the organism became so weakened that it was less resistant to the killing solution than those reared in pure water. As no deaths occurred in the culture solutions, the adaptation must have been a direct one, and in no way dependent on natural selection.

Stentor can also become acclimatised to mechanical stimuli, for Castle ‡ has observed a colony of Stentors in an aquarium being constantly struck by *Tubifex* moving backwards and forwards, and yet showing no contraction as they usually do when struck.

* For a detailed account see a paper by Davenport and Castle, Arch. f. Entwick. d. Organismen, Bd. ii. p. 227, 1895.

† Arch. f. Entwick. d. Organismen, Bd. ii. p. 564, 1896.

‡ *Vide* Davenport's "Experimental Morphology," p. 109.

Upon acclimatisation to saline solutions a considerable number of observations has been made, especially in the case of Protozoa. As long ago as 1869 Czerny * experimented on amœbæ, and found that by the very gradual addition of salt, he could acclimatise them to a 4 per cent. solution. With the unacclimatised organisms, the sudden addition of .33 per cent. of salt had in many cases a fatal effect, though some were able to stand even a 1 per cent. solution. None could resist a 2 per cent. solution, however. More recently Massart † has made quantitative determinations of the acclimatisation of certain ciliated infusoria to solutions of potassium nitrate. Unacclimatised cysts of *Vorticella nebulifera* first began to show plasmolysis when the strength of the potassium nitrate solution in which they were placed amounted to 1.2 per cent. On the other hand, cysts previously kept 22 hours in a 1.8 per cent. solution did not show any plasmolysis until the concentration was raised to 2.5 per cent. Observations on *Colpoda cucullus* gave similar results. The degree of effect produced by a .8 per cent. solution in unacclimatised organisms required a 2.5 per cent. solution in organisms previously kept 22 hours in 1.8 per cent. solution. The capacity for acclimatisation varies greatly in different organisms, for Richter ‡ succeeded in acclimatising *Tetraspora* to 16 per cent. sodium chloride solution, whilst *Spirogyra*, similarly treated, was unable to resist even a .5 per cent. solution.

The acclimatisation of certain of the metazoa to

* Arch. f. mik. Anat., Bd. v. p. 158, 1869.
† Arch. de Biol., ix. p. 515, 1899.
‡ Flora, l. p. 4, 1892.

changes of salinity appears to have been first studied by Beudant * more than eighty years ago. He placed a number of fresh water molluscs, such as *Lymnæa*, *Planorbis*, *Physa*, *Ancylus*, and *Paludina* in a vessel of water, and added a small quantity of salt every day. After a few months the water contained 4 per cent. of salt, and 170 of the original 400 molluscs were still surviving. Of another 400 kept under otherwise similar conditions in fresh water, 184 were surviving. All species are not equally adaptable, however, as *Unio* and *Anodonta*, though they throve well in fresh water, all died in salt. Beudant also performed the converse experiment of acclimatising marine molluscs to fresh water. He made observations on 38 different species of the genera *Haliotis*, *Cerithium*, *Buccinum*, *Tellina*, *Venus*, *Ostrea*, *Pecten*, and *Mytilus*. He added fresh water every day, so that after five months the animals came to live in absolutely fresh water. Out of the 38 species experimented with, 20 withstood the change perfectly well. The experiment was started with 610 individuals, and of these 375 survived. Of a similar number kept for the same length of time in normal sea water, 401 survived, or only 4.2 per cent. more. However, all the other 18 species experimented with died during the course of the experiment. In still another series of experiments, Beudant succeeded in acclimatising marine molluscs to a solution containing no less than 31 per cent. of salts. These consisted chiefly of sodium chloride, but contained also calcium and magnesium chlorides.

Numerous observations on the acclimatisation of

* Journal de Phys., lxxxiii. p. 268, 1816.

other organisms such as Myxomycetes, Actinosphærium, Crustacea, and tadpoles have been made by other observers,* but it is unnecessary to mention more than a single experiment, one made by De Varigny † upon a number of different species of marine animals. Some *Carcinus mœnas, Pagurus Prideauxii, Dromia vulgaris, Anthea cereus, Sagartia parasitica, Portunus puber, Doris tuberculata, Venus, Actinia mesembryanthemum,* and *Holothuria tubulosa* were placed in an aquarium supplied with a constant flow of water. This water was gradually diluted more and more with fresh water, with the following results:

DAY OF EXPT.	PER CENT. FRESH WATER.	ANIMALS KILLED.	DAY OF EXPT.	PER CENT. FRESH WATER.	ANIMALS KILLED.
1	22.2	nil	22	68.7	
3	33.3	nil	25	"	1 Portunus, 3 Anthea 2 Doris and 2 Venus
6	44.4	1 C. mœnas, 1 Pagurus			
11	"	1 C. mœnas, 1 Dromia and rest of Pagurus	29	77.8	
			32	"	1 Portunus, 1 Anthea
13	55.6		35	88.9	
17	"	3 Sagartia, 2 Holothuria.			

On the 38th day, when the experiment was ended, there were still living all of the original eight *Actinia mesembryanthemum* and one *Carcinus mœnas.* De Varigny suggests that the greater resistance of *A. mesembryanthemum* is probably connected with the fact that these organisms are attached to rocks near the sea

* For literature *vide* Davenport's "Experimental Morphology," p. 86.

† Centralb. f. Physiol., i. p. 566.

surface, and so are frequently uncovered. They must therefore be exposed to sea-water diluted by rivers, and to rain water. The less resistant *Anthea* is found further below the surface, however, whereas the still less resistant *Sagartia* lives in water several metres deep.

Observations on acclimatisation to saline solutions are, perhaps, less important and less interesting than those on acclimatisation to other conditions, in that, within certain limits, the phenomenon is probably a purely physical one, dependent on differences of osmosis, and the pressures and strains thereby set up. There is little doubt that if sufficient care and time be employed, any marine organism could be acclimatised to fresh water, and any fresh water form to salt water, or solutions of even greater density. If it be remembered that the osmotic pressure of a 1 per cent. solution of sodium chloride is over seven atmospheres, then it is obvious that the strain upon the tissues of an organism suddenly transferred from one solution to another of considerably greater or less salinity may easily be sufficient to rupture and kill them.

Direct observations on the acclimatisation of the vertebrata are extremely few, except in the case of certain mammals experimented on in connection with serum therapeutics. Davenport and Castle * have made some interesting observations on the acclimatisation of tadpoles to heat, however. Recently laid eggs of *Bufo lentiginosus* were divided into two lots, one of which was allowed to develop in a warm oven at a temperature of 24° to 25°, and the other kept at 15°.

* *Loc. cit.*

After four weeks, the temperature of heat rigor was determined by gradually heating the water containing the tadpoles. Whilst all the tadpoles kept at 15° went into heat rigor at or below 41°, those reared at 25° did not in any case die at a temperature below 43°, the average increase of resistance amounting to 3.2°. This adaptation to higher temperature gradually disappears on returning the tadpoles to water at ordinary temperatures, more than half of the 3.2° increase being lost after keeping them for 17 days at 15°.

Probably the capacity for acclimatisation is present to a greater or less degree in every organism. In some observations carried out at Naples,* I found that the death temperatures of a Medusa (*Rhizostoma*), a salp (*Salpa africana*), and of *Amphioxus* were, on an average, respectively 1.3°, .6°, and 1.5° higher in August than they had been in April, when of course the temperature of the sea was several degrees lower.

The adaptability of the highest organisms to changes of environment does not afford so much support to our thesis—viz., that adaptability is a fundamental property of protoplasm—as does that of the lowest organisms, because the adaptation is, as a rule, indirect and complex. Still the intrinsic interest of the subject is so great as to warrant a brief reference to it. Almost all of the exact observations deal with acclimatisation to chemical agents, especially the toxins secreted by bacteria. Upon mice Ehrlich * has made some very exact observations on adaptation to a vegetable poison, ricin. The mice were fed on food cakes soaked in

* J. Physiol., xxv. p. 131, 1899.

† Deutsche med. Wochenschr., 1891, p. 976.

solutions of the poison of increasing strengths, and after feeding for various lengths of time, the maximum amount of poison the animals could withstand was determined. This amount rapidly increased after the first day, so that after three weeks' feeding it was found to be no less than 200 to 800 times the original dose. Some of these mice were then kept on normal food for over six months, and at the end of that time could still withstand considerably more than fifty times the original amount of poison.

Even more remarkable results have been obtained in the preparation of diphtheria antitoxin. For this purpose, Roux * uses the filtrate from diphtheria bacillus cultures, it being at first mixed with an iodine solution to reduce its virulency. One-quarter cc. of the iodised toxin is injected on the first day, and this is increased to 1 cc. on the 13th day. On the 17th day ¼ cc. of the pure toxin is injected, and this is gradually increased in amount till on the 41st day 10 cc. is injected, and on the 80th day no less than 250 cc. The virulency of the last dose must have been some 5000 to 10,000 times greater than that of the first dose, and, supposing the effect produced on the horse was more or less the same after each injection, its acclimatisation to the toxin must have increased in similar proportion. As is well known, animals can be acclimatised to toxins produced by other bacteria, such as those of anthrax, tetanus, cholera, typhoid, plague, and likewise also to snake venom; but it is unnecessary to refer to these here. Upon acclimatisation in man there are probably no exact observations, but the inexact and un-

* *Vide* Crookshank's "Text-Book of Bacteriology," 1896, p. 58.

scientific are matters of common personal experience. A hot day following suddenly on a long spell of cold weather, or a cold one on a long spell of hot weather, is felt much more keenly than days of considerably higher or lower temperature which are led up to by the gradual change of the seasons. Likewise also, weather which appears very hot to one's self will be looked upon as temperate by a native Indian, or even an Anglo-Indian. Acclimatisation is often experienced by those who indulge in excessive amounts of alcohol, opium, or tobacco. For instance, De Quincey was at one time in the habit of taking 8000 drops of laudanum daily, this enormous quantity probably producing no greater effect than a dose of 30 to 50 drops in an ordinary man. Again, arsenic eaters are able to swallow as much as .4 gm. without injury, or about four times the ordinary lethal dose.

These various observations made upon members of all classes of the Animal and Vegetable Kingdoms will, I believe, be held sufficient proof of the contention that adaptability is present in all organisms, and is therefore a fundamental property of protoplasm. Whether *every* variation produced by change of environment is in the direction of adaptation to the change, it is of course impossible to say; but probably this is not the case, as, by reason of the close correlation existing between many of the characters of an organism, the change may produce a want of adaptation in some of them, but an increased adaptation in others. Supposing, however, it were possible to estimate the change produced in every character in the body, it seems to me almost certain that the sum total of all the changes would be rather in the direction of adaptation to the new surroundings, than in that of non-adaptation. It is not to be supposed for a moment that *every* one of a group of organisms exposed to new conditions of life will become better adapted to them than *any* one of the group had been originally; but merely that the characters of the group will, *on an average*, become better adapted than they had been before. Doubtless many instances can be thought of in which the effect produced by a change of environment has no appearance of being in the least adaptive, but this may be due to our ignorance of what constitutes an adaptation. For instance, it may be asked in what way a starved animal is better adapted to semi-starvation than a well-nourished one? It is, of course, less adapted in that it has, stored up in its body, less food material—such as fat and glycogen—on which it can live, but it is obviously better adapted in that its metabolism is considerably smaller than that of a well-nourished animal; *i. e.*, it actually lives on considerably less food. Again, it may be asked in what way a dusky coloured *Polyommatus phlœas* is better adapted to a warm climate than a copper-coloured one, and *vice versa* with reference to a cold climate? Possibly there is nothing adaptive about the colour of the wing scales, but doubtless it would be found that, on an average, the dusky butterflies could withstand a greater degree of heat than the coppery ones, and the coppery ones a greater degree of cold. Hence the change would, on the whole, be in the direction of adaptation.

It is probable that somatic variations, by reason of their adaptation to changed surroundings, are of very great importance in the evolution of more adaptive forms; in some cases, perhaps, of greater importance than genetic variations. Supposing, for instance, a number of organisms are more or less suddenly exposed to a considerable change of environment, whereby the majority of them are killed off. The survivors will be those which had the greatest power of adaptation to the new surroundings, and though the somatic variations will not be, as such, inherited, yet the survivors will be, on the whole, those organisms which originally possessed the largest proportion of the particular characters which have appeared as adaptive somatic variations. That is to say, adaptive somatic variations are, on an average, a magnified image of similar, but much more minute genetic variations, and hence the average hereditary characters of the survivors are in the direction of adaptation. Again, the survivors will be those individuals possessing the largest degree of innate adaptability to the particular environment in question. Hence their offspring will also possess this adaptability, and in that they will have been exposed to the changed environment throughout the whole period of development, they will show much more marked somatic variations than those shown by their parents. Finally, if it be admitted that the effects of conditions of life may be in some degree cumulative, then the adaptation of the second generation to the environment will be from this cause still further increased. Views somewhat similar to these as to the importance of somatic variations have been set forth, at considerable length and with admirable lucidity, by Professor Lloyd Morgan in his work on "Habit and Instinct" (p. 316), and to this the reader who desires more detailed discussion is referred.

Admitting that somatic variations are, on the whole, adaptive, and admitting also to a very limited extent the cumulative influence of changed conditions of life, are we to agree with Henslow * that the close adaptation of plants to their environment is due entirely to the responsive power of protoplasm to the external environmental forces, and that it is absolutely unnecessary to call in the aid of Natural Selection? By no means. Adaptive variation may be responsible for a good deal of the adaptation observed in plants, and for a very small part of that observed in animals, but probably in each case by far the larger portion must be ascribed to the ever present and ever acting agency of Natural Selection. For instance, Henslow argues very plausibly that inasmuch as certain plants when kept in a dry atmosphere develop spines and other characters similar to those possessed by desert plants, it is valid to conclude that these desert plants owe their peculiar characters to the direct action of the dry hot climate, and to that alone. Supposing this explanation to be correct, however, we ought, as Wallace points out,† to find plants with spines and the other characteristics of desert plants abounding in all dry countries, but very rare or wanting in moist and fertile districts. But this is by no means the case. Wallace states that many of the peculiarities of desert plants are present in the flora of the Brazilian Campos, and in that of the Galapagos and the Sandwich Islands, but very few of

* *Ibid.*, pp. 14 and 32.

† Nat. Science, vol. v. p. 177, 1894.

the plants indeed show any spines. Again spiny plants are exceedingly rare in the Canaries, though much of the surface, owing to long periods of drought, presents the conditions which elsewhere are supposed to produce spines. Though not prepared to deny that, other conditions equal, aridity may favour and humidity check the growth of spines, yet Wallace considers that a more important condition lies in the presence or absence of herbivorous mammals, against whose ravages the spines afford protection. Thus he mentions several countries which are not particularly arid, but in which spiny plants, and also these destructive mammals, both abound. The development of the spines is chiefly dependent, therefore, on the action of Natural Selection, and is not a direct adaptation. In other cases also Wallace believes that the "direct action of the environment can have produced only a very small portion of the modifications and adaptations that actually exist. In by far the larger number of cases no such explanation is possible, and no other adequate explanation has been suggested except variation and Natural Selection."

Though it seems to me that Wallace, by excluding all other agencies, is inclined somewhat to exaggerate the importance of Natural Selection, yet his explanation of the evolution of adaptive forms seems much more rational, and in much better agreement with facts, than that given by Henslow. The view to which the present state of our knowledge seems to me to afford best support is one which lies more or less between these two extreme explanations. It is most conveniently indicated by a diagram. Let us consider, for instance, the evolution of a typical aquatic plant from a typical terrestrial one. Supposing it were possible to estimate the extent to which characters useful to aquatic life were present in a group of terrestrial plants, and supposing we were to plot out the frequency of their distribution, then this might take the form of the curve given in the extreme left of the upper portion of the accompanying diagram. Here

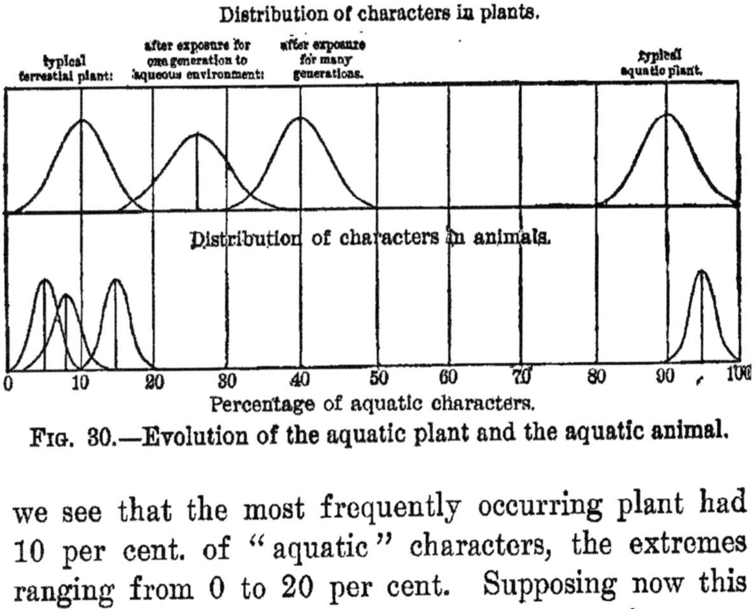

FIG. 30.—Evolution of the aquatic plant and the aquatic animal.

we see that the most frequently occurring plant had 10 per cent. of "aquatic" characters, the extremes ranging from 0 to 20 per cent. Supposing now this group of plants were exposed for one generation to an aqueous environment. It would be found at the end of that time that the proportion of aquatic characters had considerably increased, say to 26 per cent., but the frequency of distribution of the characters about the mean would still be symmetrical as it was before, the extremes now varying from 14 to 38 per cent. Some of the plants, therefore, would still possess fewer aquatic characters than were possessed by a small number of the original group of plants, in accordance with Darwin's dictum concerning plants that "whether the station (they inhabited) was unusually dry or humid, variations adapting them in a slight degree for directly opposite habits would occasionally arise." It will be noticed that in the diagram the curve of distribution of the characters is made slightly more flat topped than the other curves, indicating that the variability of a group of plants suddenly exposed to a changed environment is increased. Supposing that this group of plants is exposed to the aqueous environment for a number of generations, then, through the cumulative action of conditions of life, the adaptation will become considerably increased, and the plants will now show, on an average, say 40 per cent. of the aquatic characters of a typical aquatic plant. This increase of adaptation from the stage reached after one generation is supposed to be more or less permanent and hereditary, or would still be present if the plants were returned to their original dry land environment. But, however many generations the plants be kept in their watery surroundings, it is supposed that they will never become adapted to it like typical aquatic plants. In order to evolve such plants, Natural Selection must be present in addition, and in this case the distribution of the plants, in respect of aquatic characters, will ultimately arrive at that indicated in the curve on the extreme right of the diagram.

The lower half of the diagram is meant to represent the evolution of an aquatic animal, such as a mammal, from a land animal. Such an animal would in the first place have very few characters adapting it to an aquatic existence, and so the curve of distribution of such characters will be a more steeply sloped one than that for plants. Also the direct effect of environment in the direction of adaptation will be very much less than in the case of plants, even after exposure for a large number of generations. In fact, to effect any real and considerable change it will be essential to call in the aid of Natural Selection, and this, by acting constantly for a very large number of generations, will gradually evolve a typical aquatic mammal such as the seal, dolphin, or whale.

In spite of all that has been written to account for the almost universally present adaptation which we see in animate nature, there is still a lingering doubt in the minds of many men as to the entire adequacy of the explanations hitherto offered. It is a feeling such as this which prompted Weismann to formulate an additional principle in explanation of adaptation, and of other phenomena, as the degeneration of disused organs, viz., his theory of *Germinal Selection.** This theory supposes that, similar to the struggle for existence experienced by individual organisms, so there is a struggle among the determinants of the germ-plasm of each single individual to obtain as great a supply of nutriment as possible, and so flourish at the expense of weaker determinants. Supposing, for instance, that parts of the body, such as the hinder extremities of the quadruped ancestors of our common whales, are rendered useless. As selection ceases, individuals with

* "Ueber Germinal Selection," Jena, 1896 (English translation, Chicago, 1896).

small hind legs, represented (Weismann supposes) by weaker determinants in the germ, are as favourably placed in the struggle for existence as those with large hind legs, represented by stronger germ determinants. The weaker determinants, in their struggle with the other determinants which represent useful organs in the body, will be worsted, and gradually become more and more enfeebled, the hind legs which they represent becoming correspondingly smaller and smaller till they finally disappear altogether. Supposing, on the other hand, that the individuals showing a greater development of any particular characters than the average are for this reason favoured by Selection, then the determinants representing these characters in the germ-plasm will also be more powerful than the average, and by absorbing more nutriment will become still more robust, and produce descendants exhibiting the characters in an increased degree. That is to say, the descendants will, by Germinal Selection, become more and more adapted to the conditions in respect of which they were originally favoured by Natural Selection.

This theory, though plausible enough, is absolutely opposed to fact in so far as it relates to the evolution of more adaptive forms. As we have seen in Chapter IV., so far from the individuals selected in respect of any character tending to transmit that character in increased strength to their descendants, they almost invariably transmit less of it, or the offspring show, on an average, a greater or less degree of regression towards mediocrity, according to the amount of the character present in their more remote ancestors.

The degeneration of disused organs is, it must be admitted, a difficulty which has never been hitherto adequately accounted for, and hence, in lieu of something better, Weismann's hypothesis may in this respect be provisionally accepted. Still it is always to be remembered that it is no more than an hypothesis, which has not, and never can have, any experimental evidence to support it.

AUTHOR'S INDEX.

SUBJECT INDEX.

A

D

E

R

T